LEFT The author at Heathrow Airport.

My time at British Airways Engineering and at IAG Cargo was initially facilitated by Shona Inglis and Lewis Stagnetto from PR agency Golin Harris. These two areas are some of the airport's most fascinating, and my thanks go to Marketing and Brand Manager at BA Engineering Jonathan Harvey, Commercial Sales Manager Rajan Bindra, and Engine Health Monitoring Team Leader Trevor Ford for their support and insight.

In his role as GM of IAG Cargo's UK Operations, Mat Burton offered a great perspective on one of the most complex logistics operations in the world. My thanks also go to IAG Cargo's Global PR & Digital Manager Adam Chaudhri.

Other special mentions go to Airside Duty Manager Mark Sandford, Kevin Briscoe at Briscoe French Communications, Brian Wheeler (father of the aforementioned Brian), Beth Hurran, from the ARUP Press Office, long-time paragliding co-pilot and Emirates Boeing 777 pilot Jason Sole, ASIG turnround wizard Sajid Awan, Nick Webb, the Head of Advertising, Sponsorship & Expo, as well as Jean-Marie Lavillard, Advertising Manager at Heathrow.

Conscious it would be a sensitive subject to tackle, I received great input and support from a number of key people in the airport's security team and my thanks also go to John Walker, Head of Operations at Terminal 1. I must also acknowledge the SO18 Aviation Security team of Dawn Mordey, Chris Kerr and Inspector Dennis Bailey, who responded as quickly to my requests as they might to a real emergency at the airport.

Deep in the heart of Terminal 5, I spent time with Operations Director Susan Goldsmith and Terminal Duty Manager Mark Coleman, who together explained what it takes to run the airport's biggest terminal.

When it came to additional photos I called on long-time friend Waldo van der Waal to head to Heathrow from South Africa with his trusty bag of Canon goodies. I also kept Michael Furze and Will Kolbé from Heathrow's Photo Library busy with specific requests, and they were most obliging on every occasion. A big thank you to Chris Saunders for sharing his father's treasured album of historic Heathrow images (and for the loan of his cement mixer – but that's another story).

When the writing and photos were done, the next task was to illustrate key concepts and Roy Scorer, a talented technical illustrator from RAF Brize Norton came to the fore. Then the final chapters were passed over to Haynes, where copy editor Ian Heath, project manager Jonathan Falconer and layout wizard James Robertson set about creating the engaging spreads that follow.

Much like the subject matter itself, writing this manual was an epic journey, and my sincere thanks go to each and every one mentioned for making it possible.

Dedication

This book is dedicated to my sister Jean – for her unwavering belief in and enthusiasm for everything I have ever tackled in life. Your love and constant encouragement mean the world to me.

My sincere thanks go to my wife Tanya and daughter Hannah who have supported me every step of the way throughout this amazing journey.

Author's introduction

There's a well-known saying about 'not being able to fit a quart into a pint pot', and this is a useful analogy for Heathrow – the world's busiest international airport, which rather remarkably operates at 99.2% of its possible capacity. That means more than 1,400 flights take off and land every day from the airport's two runways (that's one every 45 seconds), making Heathrow one of the most complex operating environments anywhere in the world, where precise choreography is the name of the game. Against a backdrop of significant constraints, Heathrow has successfully adapted to take its place at the top of the aviation tree and suitably dispel the 'pint pot' analogy.

Heathrow Airport is integral to the British economy and is one of the world's most dynamic travel hubs, helping more people get to more places than any other airport in Europe. Some 80 years ago, however, it started out in life leading a relatively primitive existence, when air travel was a novelty for very few select people. Now the airport

sits at the heart of an international transportation crossroads and has developed from being little more than an infrastructure provider for airlines to an integral part of the global travel chain for both passengers and cargo. Over time Heathrow has come to realise it plays an essential role in the overall experience of 72 million annual passengers, and has adapted to deal with this challenge, delivering top-quality service despite pressure from airlines to constantly raise the benchmark.

From a personal perspective, Heathrow has played an important part in my life – it was my port of entry when I first visited the United Kingdom in 1987, and since relocating from my native South Africa in 2000 it has been my regular airport for what has been a rather exceptional amount of travel – my frequent flyer statements certainly bear testimony to this. So often am I at 36,000ft that I've narrowed my check-in time from taxi to gate to just 40 minutes and recognise many cabin crew, two of whom appear in this book. It's also not unusual for me to go shopping after a trip

abroad and feel for the brake on my supermarket trolley, thinking it's of the luggage variety.

Having always had a keen interest in aviation, my repeated trips to the airport got me thinking about how the airport actually 'works'. As a passenger I could only ever get to see a small fraction of the wider operation. It also struck me there were probably a number of people looking for the same answers. It's quite surprising, then, that nobody else had yet taken on the challenge; and so, some 18 months ago, this book was conceived.

The first of its kind, this Haynes manual provides unprecedented insight and access to every aspect of Heathrow and what makes the place tick. I've dissected Heathrow into its constituent parts, looking at everything from how the runways are constructed and maintained, and how the terminals cope with more than 200,000 passengers every day, to the complex airspace around Heathrow and, amongst other things, how the airport deals with emergencies. Most importantly, the book reflects on how all of these elements come together to ensure that Heathrow plays its vital part in what is best described as the global aviation relay race.

What has become most apparent to me during my time at Heathrow is the dedication, talent and skills of an exceptional group of committed people, many of them second or even third generation employees. They bring with them a great sense of teamwork and purpose, a fixation on safety, and a spirit that is rare in many modern organisations. Sometimes the airport makes headlines for the wrong reasons, but having got under its skin it is evident that there is always a good underlying reason for such problems as occasionally arise, coupled with a refreshing readiness to learn from the experience and to improve for the future.

The 'all areas' access I have had is granted to a limited number of people, and I consider myself most fortunate to have written this book at such a fascinating time in the airport's history. Heathrow is starting to reap the rewards of a sustained period of investment in technology, people, facilities and infrastructure, while at the same time finding itself at the heart of the debate regarding additional UK airport capacity.

I hope this book will both inform and surprise its readers, while simultaneously reaffirming the reasons why Heathrow remains one of the world's greatest airports.

Let's head backstage. Welcome aboard!

Robert Wicks
April 2014

ABOVE Heathrow's central terminal area viewed from the control tower at dusk – the airport is a constant hive of activity and one of the most complex operating environments in the world. *(HAL)*

CHAPTER 1

'Speedbird 6 inbound from Tokyo'

Speedbird 6 touches down at Heathrow. *(HAL)*

There's a gigantic series of relay races going on in the skies above us – think of the baton as an aircraft, and the runners as the various teams that interact with it along its journey between two airports, in this particular case Narita International Airport in Tokyo and Heathrow in London. The races have no start or finish – they're never-ending; but to get this particular race under way we must start somewhere, so here goes:

It's coming up for 14:00 in London, and BA006 – a British Airways (BA) Boeing 777 – is inbound to Heathrow from Tokyo with 305 people on board, consisting of 289 passengers in 4 classes, 3 pilots and 13 cabin crew. BA utilises the long-haul 777-300 for its excellent fuel efficiency and cabin comfort, principally on its routes to Asia and North America. The flight goes by the name 'Speedbird 6'. The name

'Speedbird' is the call sign used by international British Airways flights during air traffic control procedures.

The flight left Terminal 2 at Tokyo's Narita International Airport on time some ten hours earlier at 10:50 local time to commence its 5,179nm (9,591km) journey to Heathrow. The flight initially tracked north across the Sea of Japan before turning west across the vast expanses of Siberia. North of St Petersburg in Russia the flight turned south-west over Scandinavia and on to Holland before turning due west towards London.

On arrival, 65% of its passengers will be clearing customs and staying in the UK while the remaining 35% will be using Heathrow as a hub en route to other destinations. BA is by far the largest airline operator at Heathrow, with one in every two aircraft movements at the airport belonging to the national flag carrier.

Since departing Tokyo the aircraft has passed through 16 different Flight Information Regions (FIRs), with several frequency changes within each region. Think of these as the first series of baton changes in the race. The route from Tokyo to London differs on a daily basis dependent on forecast winds, weather and temporarily restricted airspace. Before taking off, the flight planning team at BA's Waterside base near Heathrow will have

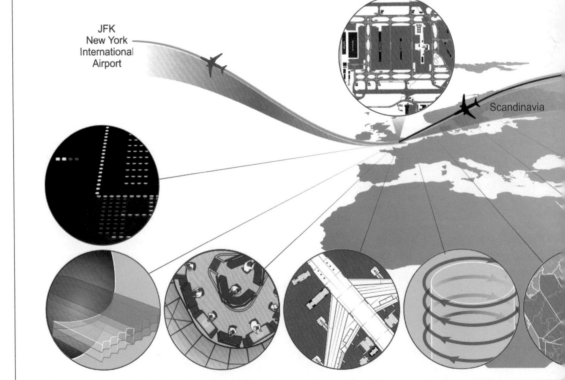

JFK
New York
International
Airport

Scandinavia

carefully studied the weather to see whether any strong jet streams might work to the advantage of the flight and will have agreed a flight plan that avoids flying into strong headwinds.

At a distance of 130 miles from Heathrow, halfway across the North Sea, between Rotterdam and Southend, the baton changes hands again as the flight is handed over from the Maastricht AAC to the London Area Control Centre (LACC), based in Swanwick in Hampshire. The captain makes contact with the London controller.

Around an hour from arrival into Heathrow the pilots request an ATIS (automatic terminal information service). This provides all the relevant airport information, such as weather conditions, which runways are in use for departure and which for landing, available approaches and any other information they might require. The ATIS helps to reduce the controllers' workload and relieve frequency congestion. The ATIS information is updated at fixed intervals or when there is a significant change, like a change of active runway.

With this information to hand the pilots brief each other on the arrival, noting any important factors such as weather, terrain, operational constraints and fuel remaining, as well as the expected taxi route to the designated stand. The pre-nominated diversion airport is also discussed,

ABOVE The distinctive tail fin designs of British Airways aircraft on stand at Terminal 5. *(Author)*

LEFT BA's Waterside Headquarters. *(Author)*

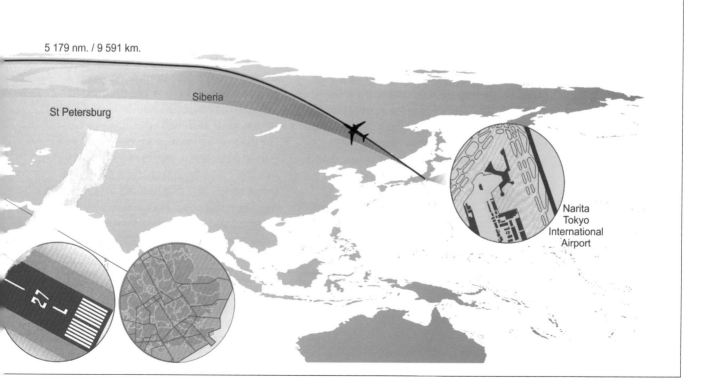

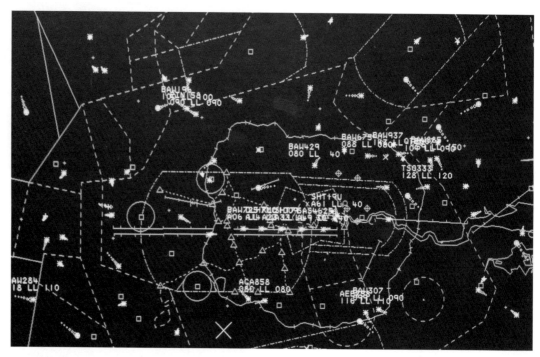

as is the weather at this location in the unlikely event the aircraft is unable to land at Heathrow. In this case fuel becomes the critical factor, but enough fuel to allow for a potential diversion will have been carried from the point of departure. The crew also discusses whether it will be a manual or automatic approach, and the procedure for a 'go-around' in the event of a missed approach. A descent checklist is then completed confirming the flap settings, speed for landing, how much reverse thrust will be used as well as the amount of auto-brake once the aircraft has touched down.

Some 125 miles from Heathrow the flight crew commences the descent. Prior to the point of descent the crew will have entered the arrival procedure cleared by ATC into the aircraft's computer, as well as the runway in use at the time. The flight crew have been informed they will be landing on runway 27L given the light prevailing westerly breeze.

At a distance of 80 miles out the baton changes hands again, and a small dot appears on the radar screens at the London Terminal Control Centre (LTCC) in Swanwick, which handles traffic flying to (or from) London's airports. Controllers at NATS (National Air Traffic Services) manage Speedbird Six's descent and any holding required before lining it up for landing into Heathrow. Soon the dot on the radar becomes a Flight Progress Strip, which is used by ATC to keep track of the flight and pass on valuable information to the next controller in the relay.

Given Heathrow's urban location, the captain is using a 'continuous descent approach' (CDA), a noise-abatement procedure designed to minimise the impact of noise on local residents. He descends at a continuous rate intended to join the final approach at the correct height for the distance. This avoids the need for extended periods of level flight, and keeps the aircraft higher for longer, using reduced thrust that in turn reduces arrival noise. Heathrow makes use of noise monitoring stations scattered around the airport that report back on excessive noise levels generated by arriving and departing aircraft.

Using ACARS (Aircraft Communications Addressing and Reporting System), a digital datalink system for the transmission of short, relatively simple messages between aircraft and ground stations, the flight crew makes contact with the airline's own control centre at Heathrow on the ground, providing them with a landing

time. By return, the crew receives details of the arrival stand at Heathrow, which in turn allows the crew to plan their taxi route and note any 'hotspots' such as possible taxiway closures. With about 40 minutes to go the pilot flying makes an announcement to the passengers, informing them of the expected time of arrival and the weather at their destination.

With 20 minutes to go and at a height of 20,000ft (6,096m), the seatbelt signs come on and the cabin crew prepares the cabin for landing. The familiar call for 'tray tables to be stowed, seats to be in the upright position with the armrest down with seatbelts securely fastened' is issued, the galley is secured, and the crew carries out a final inspection.

It's a busy afternoon at Heathrow following a strike by French ATC earlier in the day, and Speedbird 6 is directed to enter the Lambourne holding pattern for a short period before it begins the final leg of the flight.

Under the instruction of the controller on the ground, who is managing multiple flights and ensuring safe separation between them, the flight commences its approach to Heathrow at a speed of around 200kt (230mph).

On the ground, the baton passes to the BA Turnround Manager, who readies a small army of baggage handlers, refuellers, caterers and maintenance staff to prepare for the inbound flight. He will have a limited time to prepare the flight before it heads off on yet another relay.

Another vital part of the relay team on the ground is Heathrow's Airside Safety Department, whose role in the race is to ensure that their part of the 'track' – the airport infrastructure – is in tip-top condition, covering everything from the quality of the runways to keeping bird hazards at bay.

At a distance of approximately 40–50 track miles (this refers to the distance the aircraft must travel on a standard routing to approach, rather than its physical distance from the airport) the baton changes hands once again, as Speedbird 6 is handed over from London TC to Heathrow Approach, who clear it on to the airport's localiser – a vital part of Heathrow's Instrument Landing System (ILS) providing runway guidance to inbound aircraft.

Once established on the localiser, the baton is passed yet again as the flight is handed over to controllers in Heathrow's state-of-the-art ATC tower. At a distance of around two miles out on the final approach, the tower passes landing

speeds), inbound aircraft have to be sequenced by ATC at set distances apart to avoid the effects of wake turbulence.

The pilots use four stages of flaps: each stage progressively changes the shape of the wing, generating more lift, allowing the aircraft to fly at a slower speed, until eventually the final stage of flaps is selected, allowing the aircraft to decelerate to its landing speed, typically in the region of 150kt (172mph), depending on the aircraft's weight. Selection of the landing gear generates a lot of drag and further slows the aircraft; hence the landing gear is selected to the down position towards the final stages of flap selection. Once the landing gear is down and locked the speed brake is armed.

At this point the aircraft is aligned on the runway centreline using the localiser display indicator and at a suitable airspeed and rate of descent as indicated by the glide slope. The glide slope at Heathrow (and most airports) is 3°, so the aircraft flies a rate of descent based on the airspeed to give a 3° path. If the pilots exceed 3° they will be too low and will need to pitch up to reduce descent rate and vice versa. As the aircraft nears Heathrow, the airport's PAPI (precision approach path indicators) lights on the left side of the runway help the flight crew establish the correct glide path. They too are set at 3°. Visually the pilots are looking for two white and two red lights on the left and right sides of the runway respectively.

Boeing aircraft have an automatic radio altimeter that calls out as the aircraft passes through 1,000ft, 500ft, 50ft, 30ft, 20ft and 10ft.

ABOVE Passengers aboard Speedbird 6 enjoy spectacular views over London's landmarks *(Ian Black)*

RIGHT Speedbird 6 clears the localiser array. *(Author)*

clearance for the designated runway. Approaching Heathrow from the east, passengers looking out the windows can see many of central London's famous landmarks.

Ahead of Speedbird 6 and just about to touch down at Heathrow is AA6181 – another Boeing 777 – from American Airlines, inbound from Las Vegas. Not far behind the BA flight is GF003, a Gulf Airlines Airbus A330 inbound from Bahrain. With aircraft of different sizes (and approach

BELOW The flight executes a perfect landing on runway 27L. *(Nick Morrish/BA)*

BELOW Speedbird 6 taxis to its designated stand at Terminal 5B. *(Nick Morrish/BA)*

These are heights from the wheels, and just after the 30ft call out the pilot flares and lands the 220-tonne Boeing 777 in the touchdown zone on runway 27L. The spoilers extend automatically and the nose wheel touches the ground. With reverse thrust or 'idle reverse' selected, the aircraft decelerates through 60kt (70mph), and the thrust reversers are stowed once the aircraft reaches its taxiing speed.

Once Speedbird 6 has vacated the runway the baton is passed yet again, this time to one of the airport's ground movement controllers in the tower, who issues taxi instructions to the assigned stand. The pilot pushes the spoiler lever down and retracts the flaps to zero. During the taxi to stand 547 at Terminal 5B, the captain turns on the aircraft's APU (auxiliary power unit). Once the engines are shut down this will provide electrical power to the aircraft (including air conditioning for passengers) until ground power and air are connected. As the aircraft reaches its stand the parking brake is set and the engine master switches are turned off, as are the seatbelt signs.

At this point the flight plan is closed and the pilot turns off the aircraft's transponder. The flight crew completes the 'shutdown' and 'aircraft securing' checklists. Ground staff plug external power and air conditioning under the nose, allowing the flight crew to shut down the aircraft's APU. They also place chocks under the wheels as a safeguard.

Once the aircraft is safely parked the baton is firmly in the hands of the Turnround Manager, who ensures safe movement of the air bridge to the outside of the cabin door to allow the passengers to disembark. The cabin crew sort out their galleys and say goodbye to disembarking passengers.

While passengers are disembarking, the baggage and cargo teams empty the aircraft and cleaners tidy, collect rubbish and restore the cabin. In under two hours the aircraft will be turned around, refuelled with 60 tonnes of fuel, and 4.5 tonnes of catering and supplies, 3.3 tonnes of passenger bags and 21 tonnes of mail and cargo will be loaded. With nothing unusual to report from a technical perspective by the inbound flight crew, two engineers will perform a standard oil and hydraulic check, and an hour before departure a fresh flight and cabin crew will board to welcome more than 275 passengers on to what has now been transformed into BA179, heading to New York's JFK International airport.

Up in the ATC tower, the supervisor can see

ABOVE **There's not a moment to lose during the turnround procedure.** *(HAL)*

Speedbird 6 safely parked at its gate at Terminal 5, with its army of ground handlers swarming round the aircraft. 'Another one done,' he thinks to himself, but as he glances out the window to the east there's a string of lights from inbound aircraft on approach and in less than a minute the entire process will repeat itself.

In what can only be described as one of the most complex and fascinating operations in the world, flight BA006 is one of just 1,400 aircraft which land or take off at Heathrow every day. This book sets out to explain how the United Kingdom's only hub airport operates at maximum capacity, making it the world's third busiest airport and the busiest airport globally for international traffic.

In less than three hours, BA179 will be cleared to take off from Heathrow and will disappear from the screens in the Heathrow tower and reappear as a dot on the NATS screens at Swanwick, before heading west over the Atlantic.

Remember that relay race? Well, it's just started all over again....

BELOW **It's only been a matter of hours but Speedbird 6 has been transformed into Speedbird 179 bound for New York.** *(HAL)*

Chapter 2

Introduction to Heathrow

OPPOSITE Terminal 5 in silhouette at dawn as a Boeing 747-400 departs Heathrow. *(HAL)*

A brief history of Heathrow

In 1930 British aero engineer and aircraft builder Richard Fairey paid the Vicar of Harmondsworth £15,000 for a 150-acre plot to build a private airport to assemble and test aircraft. Little did Fairey know that his 'Great West Aerodrome', complete with its single grass runway and a handful of hastily erected buildings, would be the humble precursor to the world's busiest international airport.

During the Second World War the government requisitioned land in and around the ancient agricultural village of Heath Row, including Fairey's Great West Aerodrome, to build RAF Heston, a base for long-range troop-carrying aircraft bound for the Far East. An RAF control tower was constructed and a 'Star of David' pattern of six runways was laid out, offering maximum flexibility for in- and out-bound aircraft. Given the prevailing winds, two of those six original runways remain in use today.

Work demolishing Heath Row and clearing land for the runways started in 1944. However, by the time the war had ended the RAF no longer needed another aerodrome and it was officially handed over to the Air Ministry as London's new civil airport in January 1946. The first aircraft to take off from Heathrow was a converted Lancaster bomber called *Starlight* that flew to Buenos Aires.

The early passenger terminals were ex-military marquees, which formed a tented village along Bath Road. The terminals were primitive but comfortable, equipped with floral-patterned armchairs, settees and small tables containing vases of fresh flowers. Even back in those early days there was a W.H. Smith & Sons on hand. To reach aircraft parked on the apron, passengers walked over wooden deckboards to protect their footwear from the muddy airfield. There was no heating in the marquees, which meant that during winter it could be bitterly cold, but when the sun shone in summertime the marquee walls were removed to allow a cool breeze to blow through.

By the close of Heathrow's first operational year, 63,000 passengers had travelled through London's new airport. By 1951 this had risen to 796,000 and British architect Frederick Gibberd was appointed to design permanent buildings for the airport to cope with the growth in passenger numbers. His plan saw the creation of a central area that was accessed via a 'vehicular subway' running underneath the original main runway. The focal point of Gibberd's plan was a 122ft-high

ABOVE A Fairey Hendon bomber over the Great West Aerodrome at Hounslow. Note the Fairey hangar below with white writing on its roof. *(Chris Saunders)*

BELOW An early map showing the Great West Aerodrome. *(Chris Saunders)*

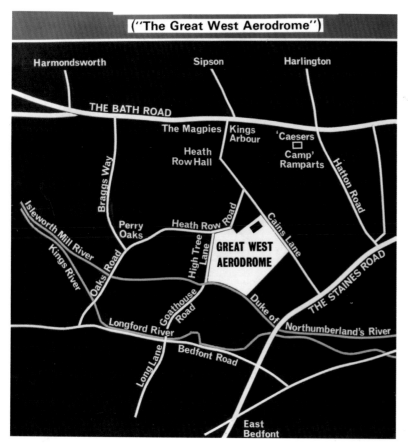

("The Great West Aerodrome")

LEFT Tents were used as terminal facilities in 1946. *(Chris Saunders/ HAL)*

RIGHT Returning passengers load their goods into a vehicle outside the terminal. *(Chris Saunders)*

BELOW The tunnel leading to the Central Terminal Area. *(Chris Saunders)*

FAR LEFT The nine-storey control tower that became operational in 1955. *(HAL)*

LEFT The interior of one of the tents – even back then, WH Smith & Son saw the benefit of a newsstand at the Terminal. *(Chris Saunders)*

control tower. There was also a passenger terminal called the Europa Building and an office block called the Queen's Building.

By 1961 the old terminal on the north side of the airfield had closed and airlines either operated from the Europa terminal (later renamed Terminal 2) or the Oceanic terminal (now Terminal 3).

Terminal 1 opened in 1969, by which time five million passengers a year were passing through the airport as the jet age arrived, with Boeing 707s, VC10s and Tridents taking travellers from Heathrow to and from all parts of the world.

The 1970s was the decade in which the world became even smaller thanks to Concorde and wide-body jets such as the Boeing 747. As the decade drew to a close 27 million passengers were using Heathrow annually. Demand for air travel also created the need for another terminal, Terminal 4, which opened in 1986.

By the time Heathrow celebrated its 60th anniversary in 2006 it had handled around 1.4 billion passengers on over 14 million flights. The start of operations at Terminal 5 in March 2008 and the new Terminal 2 development marked the beginning of an exciting new chapter for Heathrow.

The foundations on which the airport was initially built remain the same, and many of the early planning decisions have defined how the airport operates today. That said, over time the

ABOVE Terminal 2 in the 1960s showing Caravelles, Boeing 727s, DC9s and Trident aircraft. *(HAL)*

RIGHT A British Airways Concorde on stand in the 1970s. *(HAL)*

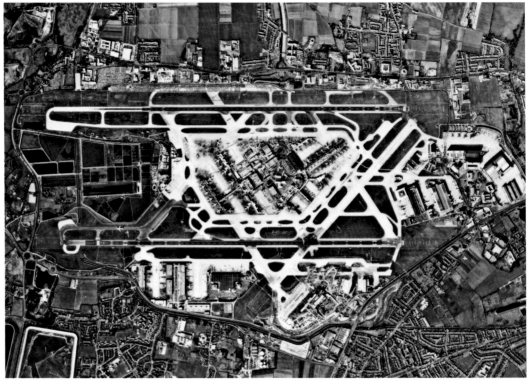

RIGHT An aerial view of Heathrow in the early 1960s. The 'Star of David' runway configuration is still evident. *(Chris Saunders)*

infrastructure has had to make way for improved facilities, as they were never originally designed to deal with the growth in passenger volume seen at Heathrow. 'The initial pattern of runways provided us a framework on which to hang the rest of our infrastructure and development for the future,' says Simon Newbold, the airport's Airside Operations Training Manager, who has been privileged enough to witness Heathrow's evolution for more than 20 years.

Today Heathrow is the world's busiest international airport and the hub of the civil aviation world. Over 72 million passengers travel through the airport annually on services offered by 90 airlines travelling to over 180 destinations in more than 90 countries.

Heathrow as a hub airport

Heathrow is the UK's only international hub. Hub airports use transfer passengers from other flights to support flights to long-haul destinations that would not be viable if they had to rely on local demand alone. This approach also allows airlines to operate flights to more destinations more frequently. Typically, passengers from short-haul flights combine with passengers from the airport's local catchment area to help fill long-haul aircraft. It is this network of flights,

transfer passengers and direct passengers that makes a hub airport different.

Operating a hub allows the UK to run direct daily flights to countries that it would not be able to sustain by itself. These flights are vital to supporting trade and economic growth. Heathrow currently serves 75 direct destinations worldwide that are not served by any other UK airport, and handles more than 80% of all long-haul passengers who come to the UK.

Hub networks play a vital role in everyday life – mail sorting offices, telephone exchanges and supermarket distribution centres are all examples which show that the best way of connecting two points efficiently is via a central hub, and Heathrow is aviation's equivalent.

To be competitive, hubs need to be able to attract network airlines and their passengers. Airlines compete with each other and will move operations to hubs that improve their profitability. This competition is good for consumers, delivering lower prices and providing greater choice of services, but – unlike its rivals in France, Germany, the Netherlands and Dubai – Heathrow is full, and its capacity constraints prevent any meaningful increase in the number of flights and routes.

The case for expanding Heathrow as the UK's sole hub airport is examined in more detail in Chapter 11.

BELOW With the exception of the two principal runways and central terminal area, the Heathrow of today bears very little resemblance to its early beginnings. *(HAL)*

Anatomy of the airport

1 Runway 09R
2 Terminal 5C
3 Terminal 5B
4 Terminal 5A
5 Runway 09L
6 Fuel Farm
7 Compass Centre
8 Fire Station
9 ATC Tower
10 Terminal 3
11 Terminal 1
12 New Terminal 2
13 Runway 27R
14 Snow Base and East
   Fire Station
15 Maintenance Area
16 Fire training Ground
17 Runway 27L
18 Terminal 4
19 VIP Suite
20 Premia Cargo
21 Ascentis Cargo
22 Airside Safety
   Department

(HAL)

Heathrow as a commercial operation

Operating an airport the size of Heathrow is a significant commercial undertaking when one considers there are multiple stakeholders as well as an extensive array of infrastructure including runways, terminals, fuel depots, hangars, fire stations and a state-of-the-art control tower. With these extensive assets and a demanding customer base, the airport must make long-term planning decisions well into the future to safeguard capacity and future business success. Amongst other things, the airport's management team – based at the Compass Centre, just to the north of the airport – has to consider capital investment, traffic forecasts, operating costs and commercial revenue opportunities, all of which have, over the past 20 years, helped to evolve Heathrow from being a simple infrastructure provider into a sophisticated and business-oriented service provider. The scale of investment at Heathrow in recent years is underlined by the £2.5 billion spent on the new Terminal 2 structure – the airport's second new terminal in recent years after a gap of over 20 years between the opening of Terminals 4 and 5.

Heathrow's annual operating costs are around £975 million, much of which is made up of daily operating costs associated with maintaining and enhancing the airport's asset base and ensuring that it can adapt to meet the needs of the world's airlines and changing trends in aviation. The airport is also required to pay extensive costs for enhanced security and the introduction of new technology. Each year around £660 million is set aside for capital investment projects to upgrade the airport.

Heathrow offsets its costs by generating income (in 2012 some £2.3 billion) from two principal streams – aeronautical and non-aeronautical revenue – made up as follows:

Aeronautical revenue	Non-aeronautical revenue
This is revenue for services or facilities directly related to the processing of aircraft, their passengers and cargo in connection with facilitating travel, including:	*This is revenue from ancillary commercial services, facilities and amenities available at the airport, including:*
Landing charges – the landing charge is based on the weight of the aircraft, including its contents, and noise (aircraft failing to meet set noise standards are subject to a higher charge). At Heathrow, a higher amount is also charged for flights landing at night.	Retail shops and services – with more than 52,000m^2 of retail space and more than 340 retail and catering outlets it's not surprising that this area is the second biggest income stream for Heathrow. Revenue is also generated from Bureaux de change services, catering, car parking and car rental services.
Departure fees – the airport levies a charge on departing passengers. More often than not this charge is paid by the airline and is bundled in the overall airfare paid by the passenger. Charges vary depending on the destination as well as the facilities and services used at the airport.	Advertising – the airport offers an opportunity for brands to interact with a highly valuable audience and Heathrow is at the forefront of airport advertising, with everything from online display advertising to experiential media, terminal campaigns and air bridge advertising.
Aircraft parking and hangar charges – these are imposed for the parking of aircraft at Heathrow and are generally based on an aircraft's weight and the duration of its stay.	Property rental – commercial income from property owned by the airport and leased to local tenants.
Other revenue streams come from non-regulated aeronautical charges, including income from airlines for air traffic services from NATS, rail services, refuelling and servicing of aircraft, use of airport property, cargo and security-related services.	

The UK's Civil Aviation Authority (CAA) regulates the amounts that may be levied by Heathrow in respect of airport charges to airlines for use of its facilities and services. This process is undertaken every five years when the airport, the CAA and the airlines develop the plans and resulting charges for the next regulatory period. The upcoming phase is Q6, and covers the period 2014/15–2018/19. Ultimately the tariffs levied by Heathrow allow the airport to recover the cost of capital, its operating costs and its capital investment net of commercial revenue.

The regulation is also designed to encourage the highest possible standards of passenger service, and comes with a series of incentives and penalties for Heathrow based on its ability to attract more passengers, reduce its operating costs, deliver higher commercial revenues and build facilities more efficiently than expected. Heathrow's success in generating income from non-aeronautical streams leads to reduced-cost airline tickets for passengers through this regulated model.

Facts and figures

- Total size of Heathrow is 1,227 hectares, or 4.7 square miles (12.17km^2).
- Heathrow has two runways and has not increased runway capacity since 1946. The northern runway measures 3,902m x 45m and the southern runway measures 3,658m x 45m.
- Number of destinations served: 184 (in more than 90 countries).
- 85 airlines operate at Heathrow.
- Heathrow has a cap of 480,000 flights per annum.

HEATHROW'S STAKEHOLDERS

Running Heathrow is a complex business and the airport's management team work in partnership with many other organisations to deliver a high-quality service. The airport has several types of clients or customers, including airlines, passengers and commercial services. The key stakeholders include:

Airlines – responsible for checking passengers in, delivering hold baggage to its final destination, cargo, providing and fuelling aircraft, boarding passengers, passenger safety and on-board catering.

The Civil Aviation Authority – the CAA controls all flight paths and aircraft routes at UK airports, and regulates airlines, airports and air traffic services. The CAA also sets airport charges at London airports.

Airport Coordination Limited – ACL is the appointed co-ordinator for London Heathrow and provides schedule facilitation services. It is the sole entity responsible for the allocation of slots at the airport. A slot allocated by ACL is effectively a permission to use the airport infrastructure.

Commercial services – the airport has to manage a wide range of individual businesses providing catering, shopping, leisure, transport and banking services across the airport.

HM Revenue & Customs – HMRC controls the import and export of goods, and prevents illegal activities such as drug, tobacco and alcohol trafficking, trade in endangered species and child pornography.

UK Border Agency – is responsible for passport control and deciding who can enter the country. This service also deals with any deportation or asylum issues.

NATS – looks after air traffic control and management, ensuring aircraft flying in UK airspace and over the eastern part of the North Atlantic are safely separated. They also provide the personnel to man and operate the Heathrow ATC tower.

Public transport operators – the airport cannot exist in isolation and depends on a massive ground transport system with several independently run bus, coach, taxi and rail companies, rental car companies, hotel shuttle services and public car park operators that provide connections to and from the airport.

- Heathrow has 133 stands served by an air bridge, 70 remote stands and 12 cargo stands.
- The number of Heathrow staff is 6,422 (August 2013), but more than 76,000 people work within the airport boundary, with a further 114,000 in work directly or indirectly related to Heathrow, creating around £5.3 billion of gross added value in the London area.
- Across the UK as a whole Heathrow supports almost 206,000 jobs, creating £9.7 billion of added value to the UK economy.
- 2013 saw just over 72 million passengers travel through Heathrow, which is an average of 200,000 passengers a day, making it the airport's busiest year on record.
- The busiest single day ever recorded was on 31 July 2011, when 233,561 passengers passed through the airport.
- 80% of the UK's long-haul flights depart and arrive from Heathrow.
- 93% of passengers are international and 7% are domestic.
- 30% of passengers are travelling for business, 70% for leisure.
- Heathrow is by far the UK's largest port of entry for overseas visitors. Around 75% of the £11 billion spent by such visitors each year is spent by those arriving by air.
- After T2 reopens in 2014 more than 60% of passengers will experience 'new' terminals,

HEATHROW VERSUS LONDON'S OTHER AIRPORTS

Based on 2012 figures.

	Runways	Terminals	Annual passengers	Annual aircraft movements	Cargo (tonnes)
Heathrow	2	5	70,037,417	475,176	1,464,390
London City	1	1	2,992,847	70,781	0
Gatwick	1	2	34,235,982	246,987	97,567
Luton	1	1	9,617,697	96,797	29,635
Stansted	1	1	17,472,699	143,511	214,160

HEATHROW VERSUS THE WORLD'S TOP TWENTY AIRPORTS

Based on 2012 figures.

Rank	Airport	Location	Passengers
1	Hartsfield-Jackson Atlanta International Airport	Atlanta, USA	95,462,867
2	Beijing Capital International Airport	Beijing, China	81,929,689
3	London Heathrow Airport	London, UK	70,037,417
4	Tokyo International Airport	Tokyo, Japan	67,788,722
5	O'Hare International Airport	Chicago, USA	67,091,391
6	Los Angeles International Airport	Los Angeles, USA	63,687,544
7	Paris Charles de Gaulle Airport	Paris, France	61,611,934
8	Dallas–Fort Worth International Airport	Dallas–Fort Worth, USA	58,591,842
9	Soekarno-Hatta International Airport	Cengkareng, Indonesia	57,730,732
10	Dubai International Airport	Dubai, United Arab Emirates	57,684,550
11	Frankfurt Airport	Frankfurt, Germany	57,520,001
12	Hong Kong International Airport	Hong Kong, China	56,064,248
13	Denver International Airport	Denver, USA	53,156,278
14	Suvarnabhumi Airport	Bang Phli, Thailand	53,002,328
15	Singapore Changi Airport	Changi, Singapore	51,181,804
16	Amsterdam Schiphol Airport	Amsterdam, Netherlands	51,035,590
17	John F. Kennedy International Airport	New York, USA	49,293,587
18	Guangzhou Baiyun International Airport	Guangzhou, China	48,548,430
19	Madrid Barajas Airport	Madrid, Spain	45,175,501
20	Atatürk International Airport	Istanbul, Turkey	44,992,420

with the other 40% experiencing the recently upgraded T4 or T3.

- More than 37% of passengers are connecting through Heathrow.
- The busiest routes out of Heathrow are New York (4.1%, or 2.9 million passengers), Dubai (2.7%, or 1.8 million passengers), Dublin (2.3%, or 1.5 million passengers), Frankfurt (2.1%, or 1.4 million passengers) and Amsterdam (2%, or 1.4 million passengers).
- In 2012, 48% of all air transport movements at Heathrow were British Airways, followed by 5% Lufthansa, 4% BMI, 3% Aer Lingus and 3% Virgin.
- Heathrow is the busiest airport in the world for international flights, with only Atlanta and Beijing boasting more flights of any sort.
- The majority of Heathrow's passenger traffic is to and from the EU (40%) and North America (23%).
- Heathrow dealt with 1,464,550 metric tonnes of cargo in 2012. Over half of that figure was cargo coming from the North Atlantic. Heathrow handles some £50 billion worth of cargo each year.
- There are currently up to 12 Airbus A380 arrivals and departures a day at Heathrow. Emirates has the most with five arrivals and five departures a day, followed by Singapore with three of each.
- The aircraft that most frequently visits Heathrow is the A320, followed by the A319.

Airport codes

Across the aviation world, Heathrow is identified by two codes:

- **LHR** – this is the airport code given to Heathrow by the International Air Transport Association (IATA). Also known as a location identifier, this code is used exclusively to designate Heathrow for public facing activities such as timetables, reservations, tickets and prominent display on luggage tags issued at the airport.

- **EGLL** – this code provided by the International Civil Aviation Organization (ICAO) applies more to industry-related activity such as air traffic control and flight planning by pilots.

IATA codes are usually derived from the name of the airport or the city it serves, while ICAO codes are distributed by region and country. In general, the first letter is allocated by continent and represents a country or group of countries within that continent. The second letter generally represents a country within that region, and the remaining two are used to identify each airport. The exception to this rule is larger countries that have single-letter country codes, where the remaining three letters identify the airport.

The Compass Centre and STAR Control Centre

So, with all of this in mind, how does the airport actually 'operate'? Before getting into the detail of runways, airfield operations and airline services, it is worth knowing a little more about two key locations.

Firstly, the Compass Centre plays host to the airport's management team. Located just to the north of the northern runway lies a linear arrangement of three large office blocks, all with a blue facade. These buildings were originally occupied by British Airways (BA) after the airline decided to consolidate its various operations into one location, but following the development of BA's dedicated Terminal 5 in 2009, as well as the airline's head office facility at Waterside (see Chapter 9), Compass House became home to what was formerly known as the British Airport Authority (BAA) before the name was changed to Heathrow Airport Holdings. It is from here that the corporate entity that is Heathrow operates, with an array of people working across core disciplines including finance, human resources, sales and marketing, press and PR.

BELOW Inside Heathrow's Star Centre – this control room sits at the very heart of daily operations at the airport. *(HAL)*

HEATHROW ID CENTRE

Another integral part of the day-to-day operations at Heathrow is the airport's ID Centre, which takes responsibility for issuing and controlling more than 100,000 ID passes.

Permanent ID passes are issued to employees and contractors who work at the airport. These passes are valid for up to five years and permit access to the areas authorised by the card on an unescorted basis. There is an extensive referencing process and criminal-records check for all Restricted Area pass applications.

Temporary passes are issued to external suppliers providing a short-term service at the airport. All ID passes are colour coded, determining the level of access granted to the holder, from 'All Areas' to 'Landside Only'.

RIGHT The Engineering Desk in the Star Centre assumes operational accountability for all of the airport's engineering assets.
(HAL)

By contrast to the substantial size of Compass Centre and situated at an undisclosed location at Heathrow is the rather intimate but 'mission critical' STAR Centre – the airport's central control room. Designed to provide a 24/7 overview and full insight into all aspects of the airport, the centre is run by the Airport Duty Manager (DMA), arguably one of the most vital roles in the airport. The centre's highly trained personnel anticipate problems to minimise the impact of disruptions to daily operations, and co-ordinate resources to respond to issues rapidly and effectively. There is not a lot that happens at Heathrow that the STAR team don't know about. The STAR Centre takes its name from the original 'Star of David' pattern of runways developed at the airport in the 1940s. Operationally, the STAR team is divided into four key areas:

1 The STAR Controllers co-ordinate any emergency response activity required by the police, fire and ambulance services. Examples range from a disruptive passenger on a flight or a road traffic accident, to more serious situations involving aircraft in distress. The operators are in direct contact with the ATC tower, which notifies them of any issues with inbound aircraft. The STAR team also provide updated information to the airport's media team, enabling them to deal effectively with enquiries. The team are trained to deal with highly complex and challenging situations, made all the more difficult today by social media channels and camera phones, which often mean that an incident is made public very quickly. A typical day can see the team receive notification of numerous incidents that need responding to either over the phone or in person. This is rather unsurprising considering Heathrow is a small city: there are more than 103,000 full ID holders on the Heathrow system, and an average of 215,000 passengers a day, plus staff and a host of other people linked in some way to the airport.

2 The Operational Monitoring Centre (OMC) provides airport management with real-time information and monitoring on the time it takes for passengers to pass through security checks and immigration before heading onward through the various terminals to their boarding gates. This allows the Airside Duty Manager (ADM) and his team to reallocate security and other resources to reduce passenger waiting times. The airport has stringent targets to meet for passenger throughput, based on the agreements it has with the airlines.

3 The third component at STAR is the Engineering desk, which assumes operational accountability for all of the airport's engineering assets, many of which are unique to Heathrow. In what effectively amounts to managing the infrastructure of a small city, a diverse array of equipment, systems and technology is in use across the airport, including hundreds of lifts and escalators, 45,000 manholes, 72 miles of high-pressure fire water mains, 81 miles of aviation fuel pipelines, and power cables carrying voltages ranging from 9V up to 400kV. The vast majority of the assets are business-critical functions, and the effective control and monitoring of these assets ensures Heathrow operates efficiently. With highly accurate and reliable data about its infrastructure, the Engineering desk is required to respond effectively to an emergency as well as to provide information about the location of existing infrastructure to the contractors. The team works on a huge range of development, construction, facilities management and maintenance projects, and with a significant capital investment programme scheduled for the coming years, engineers are more important to the airport than ever before.

Some 700 highly trained engineers respond to all major issues and provide expertise and knowledge to resolve any problems that emerge, while minor issues are typically handled by outside support teams. The engineering team has at its disposal a series of state-of-the-art software applications to monitor the airport's systems. The appropriately named Heathrow Engineering Acquisition and Remote Telemetry (HEART) system provides supervisory control of the high- and low-voltage network, the lighting substations, ventilation, the fire main pump station, sewerage plants and the airport's water treatment system, as well as Heathrow's pollution control system which is used, amongst other things,

RIGHT The Traffic Desk in the Star Centre monitors the road network in and around Heathrow. *(HAL)*

to monitor the level of glycol fluid from de-icing activities, as this has the potential to contaminate drinking water supplies and harm aquatic life. There is enough power distributed across the airport to serve a small city, and, with lighting being an essential requirement during low-visibility operations, monitoring of the power grid is vital to ensure smooth operation in all conditions. There is significant redundancy built into the system and it would take just 20 milliseconds for engineers to switch to the backup supply if this were ever necessary.

4 The fourth and final part of the STAR Centre is the Traffic desk, which monitors the road network both within and around the airport perimeter. Particular attention is paid to the tunnel system – the fourth busiest in the UK – as this is the main access route into the airport. A vehicle breakdown in the tunnel on the inbound route is likely to see a backup of traffic all the way back to the M4 motorway in just seven minutes. Traffic monitoring covers everything from road traffic accidents, road closures, diversions, car park monitoring, the de-icing of roads, lost drivers and over-height vehicles. From the STAR Centre the team can control local traffic signals to help improve the flow of vehicles across the airport's road network. Traffic management has a direct impact on passengers, staff and suppliers to the airport, so the team work hard to catch and deal with issues quickly.

From February 2014 Heathrow will be implementing an airport-wide Operations Centre entitled APOC, which will see much closer integration of the STAR Centre and the Control Centres in each of Heathrow's terminals. This new approach will enable the airport to better match capacity and demand by drawing on relevant information at the time it's needed. It brings together the various functions that provide information, making sure that the data is readily available to improve airport resilience and services to passengers and business partners. Airport Collaborative Decision Making (A-CDM) – explained in more detail in Chapter 11 – will be one of the resources contributing to this. APOC will plan, predict and proactively manage the flow of passengers, aircraft, baggage, employees and consumables using one integrated operational plan.

HEATHROW PEOPLE: MADELINE WHITE

Social Media Officer

In my role we get it all – from boy bands to lost baby bottles and the nature of the media channels means people are discussing the Heathrow brand and their experiences 24/7. It doesn't switch off, and if you work in social media you rarely switch off either.

Much of my time is spent monitoring feeds including Twitter, Facebook, YouTube, Instagram and Google+. Tweets can often be a trigger for activity in the terminal such as a celebrity arrival or flight delays. I respond to all the questions that we get asked on Facebook and this is a job that can take five minutes or fifty, depending on the complexity of the question. The questions and comments change every day. One day you could be responding to a lost baggage enquiry or car parking troubles and the next you could be pointing people in the direction of a career at Heathrow, or helping a customer who needs special assistance.

There is always so much going on at the airport and our social media channels are used to support our mainstream media activity and as a tool to support passengers. We recently worked on a story with *Top Gear*, which saw a Giallo Lamborghini Aventador given bright yellow lights to fit right in with the other 'ordinary' airport vehicles and used our channels to bring the story to life. In another instance, a passenger tweeted asking where they could watch a football match live in Terminal 3. From here we contacted the bars in T3 and found out who was showing the match, and then fed this information back to a very grateful passenger.

Recently, Heathrow teamed up with the BBC to work together to help celebrate the 50th Anniversary of *Doctor Who*. This was a perfect campaign for social media and gave us a chance to work with a brand that has a huge, international following and to bring it to life within the airport. Ultimately, we are known by our passengers – and in the social media industry – to be leading in efficient and quality responses! A career in social media didn't exist a few years ago, and now it's moving rapidly and helping to shape communication about the airport.

CHAPTER 3

Runways

OPPOSITE A Cathay Pacific Boeing 747-400 prepares to depart on runway 27R. *(HAL)*

Introduction

Heathrow's two runways are the airport's lifeblood, providing inbound and outbound aircraft with the highest-quality surfaces on which to land and take-off from. But these runways are not just extended pieces of asphalt – from an engineering perspective, Heathrow's runways are amongst the most demanding surfaces in the world, and are specially constructed to deal with the rigours of a constant stream of large, heavy jet aircraft.

In the 1940s construction began on what was, at the time, the biggest engineering project Britain had ever seen – no other airport in the UK had been built to the same scale. Writing in his book *Time Flies: The Heathrow Story*, Alan Gallop says: 'Daily convoys of buses and lorries collected runway gangs from town squares, railway and underground stations and drove them to the giant building site.' During the first seven months of construction on Heathrow's runways, some two million tonnes of earth were excavated. Once the first runway was built, a taxiway and apron was laid adjacent to Bath Road, and this was followed by the construction of two further runways. Gallop concludes: 'All three runways were built with a view to being extended as soon as the Air Ministry saw fit to do so … and by Christmas 1945, enough concrete had been mixed and poured onto the runways to build a new road from London to Edinburgh. Now all that was needed was a peacetime commercial aeroplane to christen it.'

By the 1950s, with demand for air travel increasing, three runways had become six, in the distinctive 'Star of David' pattern, the greatest of which was 3,000yd long and 100yd wide. In addition to the two current east-west runways, four runways ran diagonally across the site, giving aircraft a host of options depending on the wind direction. Over time it became clear that the prevailing wind direction was from the west, and today modern aircraft can take off and land on the north and south runways in all wind conditions and the cross runways are no longer required, so that some 70 years on there is little evidence of their former existence.

Keeping the runways fully operational is one of the highest priority tasks at the airport. Closures, even for short periods, have a significant knock-on effect and can mean aircraft are diverted to other airports, reducing income for Heathrow and inconveniencing passengers. With a current cap of 480,000 annual flight movements, equating to some 1,400 aircraft movements every day, Heathrow's two runways are at maximum capacity and there is a pressing need to increase capacity at the country's only hub airport.

Layout and dimensions

For normal fixed-wing aircraft it is advantageous to perform take-offs and landings into the wind, and as a result runway orientations are decided based on historical wind patterns and directions. If the winds are more variable in direction and the airport is large enough to financially justify the investment, airports can have several runways in different directions, so that a runway can be selected that is most closely aligned with the wind.

In theory, Heathrow Airport has four runways, which go by the designations of 09L, 09R, 27L and 27R; but physically there are only two runways – 09L/27R (the northerly runway) and 09R/27L (the southerly). Both point in the same east–west direction and are classed as parallel runways. Currently the remains of the original runway configuration can still be seen on the airfield, but with ongoing development at the airport these will soon disappear almost completely from sight.

The runways are 50m wide for the contact lane plus an additional 20.5m shoulder on each side to allow for jet blast. The northern runway is 3,902m long, while the southern runway is marginally shorter at 3,660m. The distance between the two runways is 1,414m. To put these dimensions into perspective, each runway could accommodate more than 30 full-size football pitches.

Like all major airports, Heathrow is required to publish the vital distances for take-off or landing, which are used by pilots to calculate the take-off and landing performance of their aircraft. These numbers may change in the unlikely event of a surface defect or incident. Declared distances are based on the following:

- TORA (take-off run available) – refers to the length of runway declared available and suitable for the ground run of an aircraft taking off.
- TODA (take-off distance available) – the length of the take-off run available plus the length of

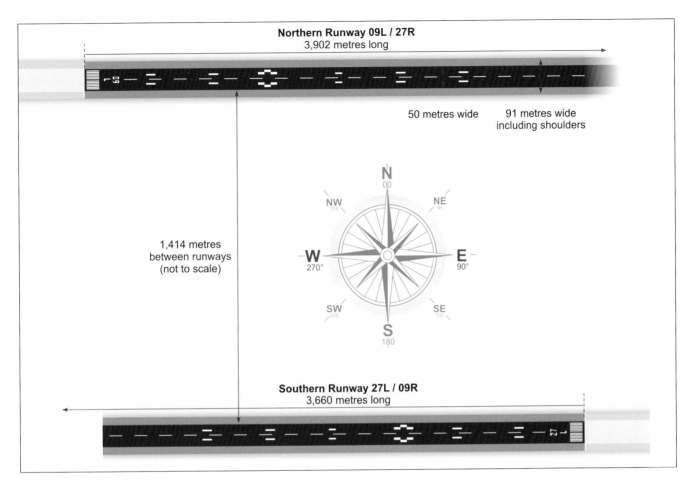

50 metres wide

91 metres wide
including shoulders

1,414 metres
between runways
(not to scale)

N
00

NW
315

NE
45

W
270°

E
90°

SW
225

SE
135

S
180

Southern Runway 27L / 09R
3,660 metres long

the clearway, if provided (an area at the end of TORA which does not have to be paved, but is clear of obstacles, over which an aircraft may make a portion of its initial climb).

■ ASDA (accelerate stop distance available) – the length of the take-off run available plus the length of the stopway (a rectangular area at the end of the TORA designated as a suitable area within which an aircraft can be stopped in the case of an aborted take-off) if provided.

■ LDA (landing distance available) – refers to the length of runway that is declared available and suitable for the ground run of an aircraft landing.

The current declared distances for Heathrow are:

Runway	TORA	TODA	ASDA	LDA
09L	3,902m	3,902m	3,902m	3,595m
27R	3,884m	3,962m	3,902m	3,902m
09R	3,660m	3,660m	3,660m	3,353m
27L	3,660m	3,660m	3,660m	3,660m

Safeguards

Heathrow Airport is officially safeguarded by the CAA with a series of procedures and policies to protect aircraft in flight and on the ground, to protect visual and radio aids, and to protect people on the ground. Outside of the airport the main concerns are the height of new structures or a major attraction to birds within a radius of eight miles (13km) of Heathrow.

According to official statistics 80% of aircraft accidents occur during take-off and landing. As a result, Public Safety Zones (PSZs) have been established at the ends of the runways. The PSZ rules prevent a significant increase in the number of people residing, working or congregating in the area. A PSZ broadly coincides with the shape of the approach surface and extends to just under 1.5km.

Visual aids, such as approach lights, are protected from being obscured, and the CAA has powers over the display of lights in the area which might be dangerous or confusing. Radio aids for navigation and radar have restricted areas and invisible safeguarded surfaces which must not

ABOVE **Schematic of Heathrow's two parallel runways.** *(Roy Scorer)*

be infringed, as some could be distorted or be susceptible to reflection.

Aircraft taking off and landing are protected by three-dimensional blocks of airspace, the lower limits of which form a series of obstacle limitation surfaces summarised as follows:

- **Conical surface** – a 1:20 slope to a height of 150m from the end of the inner horizontal surface.
- **Outer horizontal surface** – extends from the end of the conical surface to a distance of 15km from the midpoint of each runway.
- **Inner horizontal surface** – 45m in height x 4,000m radius.
- **Transitional surface** – 1:7 slope from the edge of the runway strip.
- **Approach surface** – 1:50 and 1:40 slope.
- **Take-off climb surface (TOCS)** – 1:50 slope.
- **Obstruction-free zone (OFZ)** – a set of inner obstacle limitation surfaces for precision approach runways consisting of inner approach surface, inner transitional surface and abort landing surface (go-around).

BELOW Various safeguards are in place to protect aircraft, people and critical safety assets on the ground. *(Roy Scorer)*

Construction

From an engineering perspective, runways for large jet aircraft are amongst the most demanding surfaces in the world, and with the volume of flights that Heathrow deals with on a daily basis the construction and ongoing maintenance of the runways is of paramount importance.

Runway pavement surfaces are made and maintained to maximise friction for the best possible braking performance of inbound aircraft. The surface needs to be homogenous to avoid large expanses of rainwater and ice, and, even more importantly, to withstand the heavy loading of arriving and departing aircraft. An Airbus A380 weighs about 380 tonnes and lands at a speed of almost 200mph. At take-off the aircraft can weigh as much as 560 tonnes.

Aircraft manufacturers spend considerable time designing landing gear to support the weight of their planes on larger and more numerous tyres, thereby minimising stresses and adverse effects on the runway surface. The A380 undercarriage, for example, consists of 22 wheels – four main landing

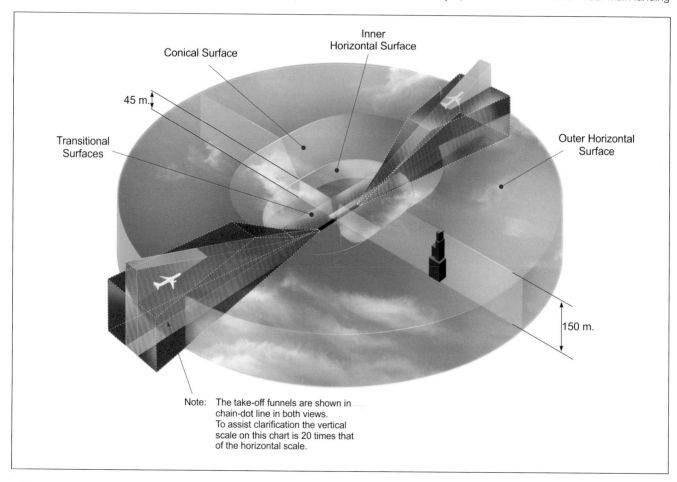

Conical Surface

Inner Horizontal Surface

45 m.

Transitional Surfaces

Outer Horizontal Surface

150 m.

Note: The take-off funnels are shown in chain-dot line in both views.
To assist clarification the vertical scale on this chart is 20 times that of the horizontal scale.

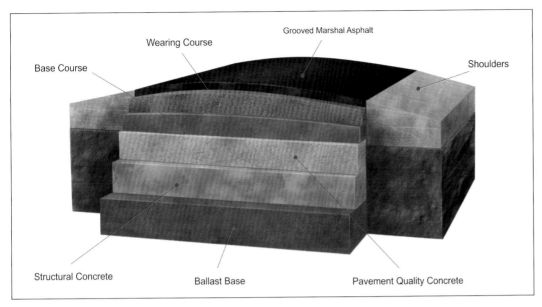

Base Course

Wearing Course

Grooved Marshal Asphalt

Shoulders

Structural Concrete

Ballast Base

Pavement Quality Concrete

gear legs and one nose leg, with the two inboard landing gear legs each supporting six wheels.

Heathrow's original runways were constructed to a width of 91m (300ft), largely in anticipation of the expected development in aviation that would see aircraft such as the Brabazon take to the skies. The Bristol Type 167 Brabazon was a large, propeller-driven airliner designed by the Bristol Aeroplane Company to fly transatlantic routes between the United Kingdom and the United States. The first prototype was flown in 1949, only to prove a commercial failure when airlines felt the airliner was too large and expensive to be useful. Despite its size it was designed to carry only 100 passengers, albeit in roomy conditions not generally found in modern aircraft. In the end only a single prototype was ever flown, but the initial design of the airport's runway and planning has proved most fortuitous as the size of aircraft has progressively increased.

The airport's runways are constructed of several layers, which start with a ballast base, a layer of structural concrete and a layer of pavement-quality concrete. This is followed by a layer of base course, and then a layer of wearing course beneath the top layer of Grooved Marshall Asphalt, which is what is visible to the naked eye. This is designed specifically to withstand the very high load and shear forces from landing aircraft.

The runways have to be prepared and maintained to maximise friction for wheel braking. Good friction characteristics are achieved by a combination of a curved surface to provide a slope for efficient surface water run-off, and surface texture to increase friction and allow

water to disperse from the tyre contact area. The cambered Grooved Marshall Asphalt on both runways at Heathrow allows water to drain on to the transverse slope either side of the crown of the runway, on to the shoulders themselves, and eventually down storm drains. With the surface water flowing off through the grooves, the peaks between the grooves give aircraft tyres a suitable contact surface.

The design and maintenance of the runways and taxiways at Heathrow is based on the ACN and PCN systems. Once constructed, the strength of a runway is measured and receives a pavement classification number (PCN) from the International Civil Aviation Organization (ICAO). The PCN is used in combination with the aircraft classification number (ACN), which indicates the relative effect of the load a particular aircraft exerts on the runway. The PCN refers not only to the runway but also to the aprons and taxiways – while

an aircraft is being loaded and taxiing prior to departure the apron experiences significant loads from the aircraft, so these areas have to be of comparable strength. Provided the PCN is equal or greater than the ACN, the use of the pavement is unlimited.

Heathrow's current design PCN is 083RAWT, one of the highest available ratings, which is broken down as follows:

083 The numerical value indicates the load-carrying capacity of the pavement and is calculated based on a number of factors, such as aircraft geometry and the runway's traffic patterns rather than simply being an expression of the direct bearing strength.

R Refers to the fact that Heathrow has a rigid, rather than flexible, design.

A An expression of the strength of what is underneath the pavement section, known as the sub-grade, so Heathrow's is deemed to be very strong.

W An expression of the tyre pressures permitted (W indicates tyres of any pressure).

T Confirms that the value was achieved by technical evaluation (as opposed to a physical testing regime).

Components and markings

Each runway at Heathrow consists of a number of integral components and markings, including:

■ **Threshold** – this describes the start of the runway available and suitable for the landing of aircraft. At the eastern end of the runways the threshold is almost at the start of the pavement; however, at the western end Heathrow has displaced thresholds (due to height restrictions on the approach paths).

■ **Threshold markings** – sometimes referred to as the 'piano keys', these mark the start of the threshold.

■ **Touchdown zone markers** – these parallel stripes indicate the target area on which pilots should put the wheels of their aircraft when landing. Aircraft touch down about 300m–400m into the runway, but to allow for variations the first 900m is marked by paint and lights. These are 400m from the threshold.

■ **Aiming point** – located forward of the centre of the touchdown zone (TDZ). It is positioned so as to compensate for the angle of approach from a pilot's point of view. It is intended to be used as a visual reference by the pilot and corresponds to where the glide slope intercepts the runway surface. The Instrument Landing System (ILS) guides aircraft to these points.

■ **Centreline** – this identifies the centre of the runway and provides alignment guidance during take-off and landings. It consists of a line of uniformly spaced stripes and gaps that run the full length of the runway.

■ **Windsleeves** – there are four illuminated windsleeves located close to the thresholds of each of the main runways. It is an ICAO requirement for the pilot to have a visual reference of the wind speed and direction on the airfield.

■ **Shoulders** – these concrete areas to the side of the contact lines on the runways are capable of supporting aircraft without causing structural damage and are there to prevent erosion by blast from overhanging engines.

■ **Blast pads** – these can be found at the end of each of the runways and are designed to prevent erosion caused by aircraft running their engines up to full power just before they start their take-off runs. They are marked by large yellow arrows at the eastern end, and although not marked they are included in the displaced threshold area at the western ends.

BELOW Runway markings and measurements.
(Roy Scorer)

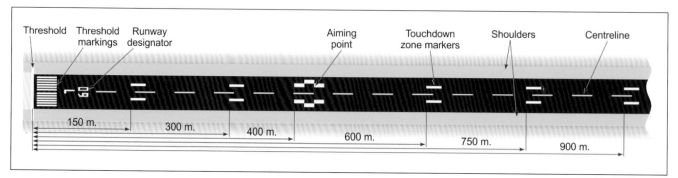

- **Clear and graded area (CGA)** – this is an area within the runway strip 105m wide from the runway centreline narrowing to 75m at the ends, where the ground is prepared in such a way that, if an aircraft runs off, it will not sustain significant damage.
- **Runway designators** – the runways are identified by a painted designator at the start of the runway. As Heathrow has two parallel runways which are used from both ends, a letter is added to distinguish them further, hence the markings 27R/09L for the northern runway and 27L/09R for the southern runway.

Navigational aids

There are several navigational aids at Heathrow that are essential to safe runway operations, including:

Instrument Landing System (ILS)

Aircraft inbound to Heathrow are sequenced for safe separation by controlling the speed and lengths of routings prior to the aircraft being turned on to their final approach. Both runways at Heathrow are precision approach runways, which means a fully calibrated Instrument Landing System (ILS) is provided at the airport. The ILS consists of a localiser aerial to guide aircraft down the centreline of the runway and a glide-path aerial to guide aircraft on to the correct angle of approach. The localiser aerial in use is at the far end of the runway, whereas the glide-path aerial is to the side of the runway, and approximately 300m upwind of the threshold.

Both the glide-path aerial and the localiser aerial are protected areas, incursion into which may bend the ILS signal and create a dangerous situation by giving spurious readings to an approaching aircraft.

The beam has a protected range that extends out a horizontal distance of 25nm. The glide slope at Heathrow is set at 3°, which is the angle recommended by the International Civil Aviation Organisation (ICAO) for commercial aviation for safety reasons. Steeper angles are generally only accepted if required to avoid obstacles. The glide slope gives information to the pilots, informing them where the touchdown point is on the runway, giving the aircraft the correct descent profile. Distance measuring equipment (DME) gives information to the aircraft regarding its distance from the runway touchdown zone, and is located approximately at the midpoints of the runways.

HOW RUNWAYS ARE NAMED

The large painted number at the end of each runway is called its designator. Runways around the world are designated by a number between 01 and 36, which is generally one-tenth of the magnetic azimuth of the runway's heading. In simple terms, a runway numbered 09 points east (90°), runway 18 points south (180°), runway 27 points west (270°) and runway 36 points to the north (360° rather than 0°). Since runways can be normally used in two directions – as is the case at Heathrow – they have a second number which will always differ by 18 (180° or halfway across the compass) given that it is in the directly opposite position.

Tokyo's Narita Airport, for example, also has two parallel runways designated 16L/34R and 16R/34L with runways running at 160° (towards the south-east) and 340° (towards the north-west).

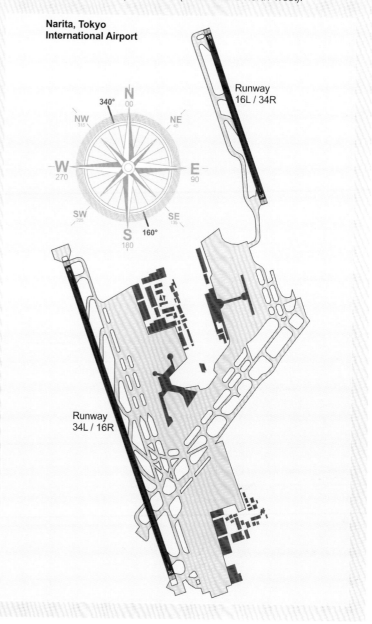

Narita, Tokyo International Airport

Runway 16L / 34R

Runway 34L / 16R

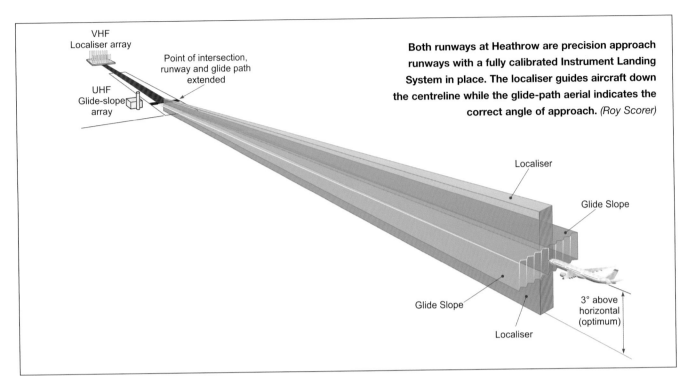

VHF Localiser array

Point of intersection, runway and glide path extended

UHF Glide-slope array

Both runways at Heathrow are precision approach runways with a fully calibrated Instrument Landing System in place. The localiser guides aircraft down the centreline while the glide-path aerial indicates the correct angle of approach. *(Roy Scorer)*

Localiser

Glide Slope

Glide Slope

Localiser

3° above horizontal (optimum)

RIGHT The ILS localiser array. *(Author)*

The glide slope transmitter is placed at the end of the runway that the aircraft is landing on, whereas the localiser transmitter, which projects the localiser signal, is located at the opposite end of the runway to allow the aircraft to be guided along the full length of the runway.

As a result of the set 3° approach angle, aircraft using the ILS will be at a set height for distance

CAT II AND III OPERATIONS

At Heathrow there are times when a visual approach and landing are undertaken – it may be that the ILS is switched off or happens to be unserviceable. At Heathrow the ability to land in lower visibility is required, so a CAT III B Precision Instrument Approach runway is provided when visibility is down to 50m and there is cloud at ground level.

A non-instrument or visual runway has no radio-based landing aids and the pilot judges his landing from what he can see. Runway lighting, including approach lighting, is needed if the runway is to be used at night. Good visibility is required for operations on such a runway.

CAT III refers to a low-visibility instrument landing procedure for aircraft and pilots that are certified and qualified to perform them. CAT III and CAT II approaches require the use of an autoland system for the aircraft, which essentially means that the autopilot lands the aircraft while the pilots monitor the aircraft and autopilot closely for any detected anomalies during the approach. CAT III landings on the most modern aircraft do not require the pilots to see the runway prior to landing. CAT II landings typically require the pilots to be able to see the 'runway environment' by 100ft above the landing threshold by either a radio altimeter or barometric altimeter, depending on the airport. If the runway environment is not in sight (CAT II), or any anomalies are detected during the approach, a 'missed approach' (called a 'go-around' – see page 44) is performed.

RIGHT Runway 27L CAT II/III. *(Author)*

from the runway. At the same distance, each aircraft will be at about the same height. With this information readily to hand, inbound pilots and their aircraft can find themselves on a heading which is the same as the runway centreline; they will know how far they have to travel and the angle at which they need to travel until they are over the touchdown zone.

Landing is, of course, a very busy stage of flight and it is essential for safety reasons that aircraft are stabilised in their approach some distance from touchdown. This means that, amongst other things, an aircraft is set up for landing and has its approach speed stabilised.

In the interests of separation and safety, Heathrow's air traffic controllers have discretion over where they direct aircraft to join the ILS in order to allow aircraft to be stabilised at an appropriate distance from touchdown. It sounds fairly complex, but the system is so accurate it can guide 300 tonnes of metal down a line just 15m in width from a height of 2,500ft and a distance of eight miles on to a touchdown zone and straight down the middle of the runway.

Primary and secondary surveillance radars

Primary radar works on a response basis, whereas secondary radar works on an interrogation basis. Both look after approach and airborne movements.

Surface movement radars

These provide actual ground movement information of aircraft and vehicles to ATC. Working in conjunction with these are another 15 sites around Heathrow, providing additional point location information.

Transmissometers

There are three transmissometers for each runway. These are used to measure Instrumented Runway Visual Range (IRVR), which is used as one of the main criteria for minima on instrument approaches, as in most cases a pilot must obtain visual reference of the runway to land an aircraft.

Surface wind display systems (SWDS)

These anemometers are located in the ILS glide-path areas and provide surface wind speed and direction information to ATC that is in turn passed on to pilots of approaching or departing aircraft.

Runway operations

Of all the challenges that Heathrow faces, noise associated with aircraft movements is arguably the most significant. Despite the challenges of noise during the day and night from arrivals, departures, go-around activity, ground movements and the four stacks around London, Heathrow is committed to enhancing the

ABOVE LEFT A secondary surveillance radar operates at Heathrow looking after approach and airborne movements. It works on an interrogation basis, has a 60-mile range and is Heathrow specific. It is manufactured by Raytheon and is located on the south side of the airfield to the east of Terminal 4. *(Author)*

positive and minimising the negative impacts of the airport's operations on the environment. To this end various noise abatement procedures are followed and these have a direct bearing on how Heathrow's two runways operate.

Arrivals

Unlike departing aircraft, there are no fixed routes between the holding stacks and the ILS leading to the final approach for arriving aircraft; neither are there any noise limits or fixed heights. This is because arriving aircraft approach UK airspace in a random pattern and then have to be sequenced to ensure safe separation. On average approximately 70% of arrivals come from the east (over London) and about 30% from the west (over Windsor). As it is due to wind direction, the direction of operation cannot be controlled or predicted. Heathrow is not subject to restrictions from the government in terms of noise limits for arriving aircraft but has taken a host of pro-active stances by introducing various abatement procedures to minimise the impact of arriving aircraft on both the local and wider community. These include:

Runway alternation

First introduced in the 1970s, a system of runway alternation is in place for landing aircraft during westerly operations (*ie* when aircraft arrive over London), and offers predictable periods of relief from the noise of landing aircraft for communities under the final approach tracks to the east of the airport. One runway is used for arrivals and one for departures. This then alternates at 15:00 local time to provide a respite for those directly under the flight path.

Continuous descent approach

One of the main noise abatement measures is continuous descent approach (CDA), a technique of flight during which a pilot descends at a rate aimed at achieving a continuous descent to join the glide-path at the correct height for the distance. This procedure thereby avoids the need for extended periods of level flight. The intention of a CDA is to keep aircraft higher for longer, using reduced thrust and thereby reducing arrival noise.

LEFT Virgin Atlantic Boeing 747-400 'Tubular Belle' touches down on runway 27L. *(Author)*

'Joining point rules'

These dictate at which point aircraft must be established on the ILS before being able to descend further. At night the requirement is 500ft higher than during the daytime, keeping aircraft higher for longer.

Limiting the use of reverse thrust

Pilots are asked not to use reverse thrust when landing between 23:00 and 06:00, unless they need to for safety reasons. This helps to reduce disturbance in areas close to Heathrow.

Financial incentives

Heathrow charges noisier aircraft more to land than quieter aircraft. This acts as an incentive to airlines to introduce quieter fleets.

Departures

The airport has to deal with approximately 650 departures every day. Approximately 70% of take-offs head toward the west (over Windsor) and about 30% to the east (over London). As it is due to wind direction, the direction of operation cannot be controlled or predicted. Aircraft leaving Heathrow are required to follow flight paths known as noise preferential routes (NPRs) up to an altitude of 4,000ft. NPRs were set by the Department for Transport (DfT) in the 1960s and were designed to avoid overflight of built-up areas where possible. Each NPR has a 'swathe' measuring 1.5km either side of the route centreline, resulting in a corridor 3km wide. As long as the aircraft are within this 'swathe' they are considered to be on track. They do not have to follow the centreline of the NPR. Other important points to note are:

The Cranford Agreement

This was a verbal agreement, dating from the 1950s, to avoid use of the northern runway for take-offs in an easterly direction over the village of Cranford unless necessary for operational reasons or essential maintenance. Therefore aircraft depart to the east using the southern runway and arrive using the northern runway. The agreement has now come to an end, thereby allowing departing aircraft to use the northern runway (09L) on easterly operations, which in turn will allow the introduction of runway alternation for arriving aircraft on easterly operations, redistributing noise more equitably around the airport. However, because Heathrow has developed within the context of the Cranford Agreement it's not yet geared up to full-time runway alternation – there

are simply too few access taxiways to the northern runway and too few exit taxiways from the southern runway to achieve an adequate movement rate to meet demand, particularly at peak times. Rapid exit taxiways (RETs) are linked to runways at an angle (ideally 30°), allowing aircraft to vacate the runway at higher speeds, which in turn permits another aircraft to land or depart in a shorter space of time, thereby improving overall efficiency. As part of the current refurbishment programme, and to take into account the end of the Cranford Agreement, Heathrow has had to apply for planning permission to provide the additional RETs for the southern runway and changes to the access point for departures on the northern runway, which it aims to complete during 2014.

'1,000ft rule'

After take-off, aeroplanes are required to climb to at least 1,000ft above the airport level by 6.5km from the point when they begin moving on the runway. This encourages aircraft operators to gain height as quickly as possible and then reduce engine power and noise at the earliest opportunity.

ABOVE A Cathay Pacific Airbus A340 comes in to land with landing lights in foreground. *(HAL)*

BELOW Photo sequence of an Aer Lingus Airbus A320 departing runway 27R. *(Waldo van der Waal)*

Fines

Noise generated by departures is continually monitored, and if an aircraft creates more than is allowed the airline is fined. Money collected is distributed to a wide range of community projects in areas affected by the airport's operations.

Aircraft noise on the ground

People living in close proximity to the airport are likely to hear noise from aircraft on the ground. Sources include aircraft using reverse thrust when landing, aircraft moving between runways and stands, ground power units and engine testings. Unlike the noise limits that apply to departing aircraft, there are no limits on other sources of noise that originate from the airport; however, the airport has various controls in place to balance the interests of the local community and the needs of airport users.

Stacks

Heathrow's four stacks (see Chapter 4 for information) are where aircraft are held while they wait to land. The minimum height of aircraft in the stack is 7,000ft (2,133m), and although their noise should not cause a nuisance on the ground people in the countryside can be affected, since there is often very little background noise and the aircraft noise consequently becomes more noticeable. People living between the stack and the final approach may hear noise as the aeroplanes leave the stack and make their way to the final approach to Heathrow.

Night noise

The noise created by aircraft at night is thought to cause more disturbance to some people because there is less background noise from other sources, and the majority of people will be trying to sleep. Heathrow has always been a 24-hour-operation airport and there has never been a night ban; however, the airport's management try to balance the interests of local communities and those of the airport's users through restrictions on night flights put in place by the DfT. Night-flying restrictions are divided into summer and winter seasons. These consist of a movement limit and a quota count system. This means that points are allocated to different aircraft types according to how noisy they are. The noisier the aircraft type, the higher the points allocated, providing an incentive for airlines to use quieter aircraft types.

Go-around activity

'Go-around' is a procedure adopted when an arriving aircraft on final approach aborts landing by applying take-off power and climbing away from the airport. It is a set procedure to be followed by the flight crew in the event of an aircraft being unable to land. Go-arounds are flown in the interests of safety and typically account for less than 0.25% of total arrivals at the airport.

Taxiways

Heathrow's taxiways are the airport's 'road network for aircraft' and create efficient routes to and from the runways to the terminal gates. There are some 18 miles of taxiways at Heathrow and they are named using a system of phonetic and numerical illuminated sign boards that conform to international ICAO standards.

Aircraft at Heathrow are controlled on the ground using the taxiway designation system. The route given to them includes the taxiways to be followed and any intermediate holding positions. For example, an aircraft vacating runway 27L might be given instructions along the following lines: 'Straight ahead Echo, hold Echo 2.' The aircraft will be transferred to the next ground movement controller (GMC), who will then instruct: 'Right Bravo, Juliet, Stand 117.'

The width of taxiways depends on the size of aircraft expected to use them, particularly taking into consideration the width of their main undercarriage (known as 'wheel track'). Taxiways are widened at junctions and bends – known as 'fillets' – to allow the main undercarriage of aircraft to manoeuvre as it follows the curve.

The construction of Heathrow's taxiways varies depending on their age and the rehabilitation work carried out on them. They principally consist of concrete bays separated by expansion gaps, filled with joint seal. Many areas of taxiway are also covered with Marshall Asphalt. Even though

BELOW An Emirates Airbus A380 heading for Dubai departs from the southern runway. *(Author)*

taxiways are designed to an internationally agreed standard of width, this still presents problems to grass areas where the outer engines of larger aircraft overhang. Therefore to ensure that aircraft can safely manoeuvre and avoid any obstacles, an area on either side of the taxiways is provided that is clear of objects that may endanger taxiing aeroplanes. This is referred to as the 'taxiway strip and graded area'.

Signage

There are multiple illuminated signs around the airfield to afford additional operational guidance and location information to pilots and vehicle drivers on the airfield, particularly in conditions of low visibility.

With the introduction of the Heathrow taxiway designation system in the mid-2000s, a number of additional illuminated sign boards were introduced as reporting points to assist ATC in directing movement on the airfield, and as points at which pilots are handed over to another controller. These signs are generally located at the entry to the runway holding areas and are as follows:

Runway reporting points				
27R	Pluto	Satun	Titan	AY1, AY3, AY4, AY5
27L	Lokki	Ettiv	Morra	Nevis
09R	Oster	Horka	Vikas	Dasso	Hanli
09L	Snapa	Rabit	Dingo	Cobra

Mandatory instruction signs have a red background with white inscriptions and are provided in order to identify runway holding areas where an aircraft or vehicle must not proceed without ATC clearance. At the end of the main runways where an aircraft is liable to start a take-off run the ILS category is also added – for example, '27R CAT II/III'.

- Taxiway location and direction information signs are combined in the same sign, but destination information is provided in a separate sign.
- Location is indicated by displaying the taxiway designation letter in yellow on a black background surrounded by a yellow border. For example, to indicate the approaching end of a taxiway a diagonal yellow line is added.
- Direction or destination information is displayed with black lettering on a yellow background and is situated at the approach to a junction.

ABOVE Aerial view of aircraft using Heathrow's taxiway system. *(HAL)*

LEFT Signage across the airfield provides guidance and location information to pilots. *(Author)*

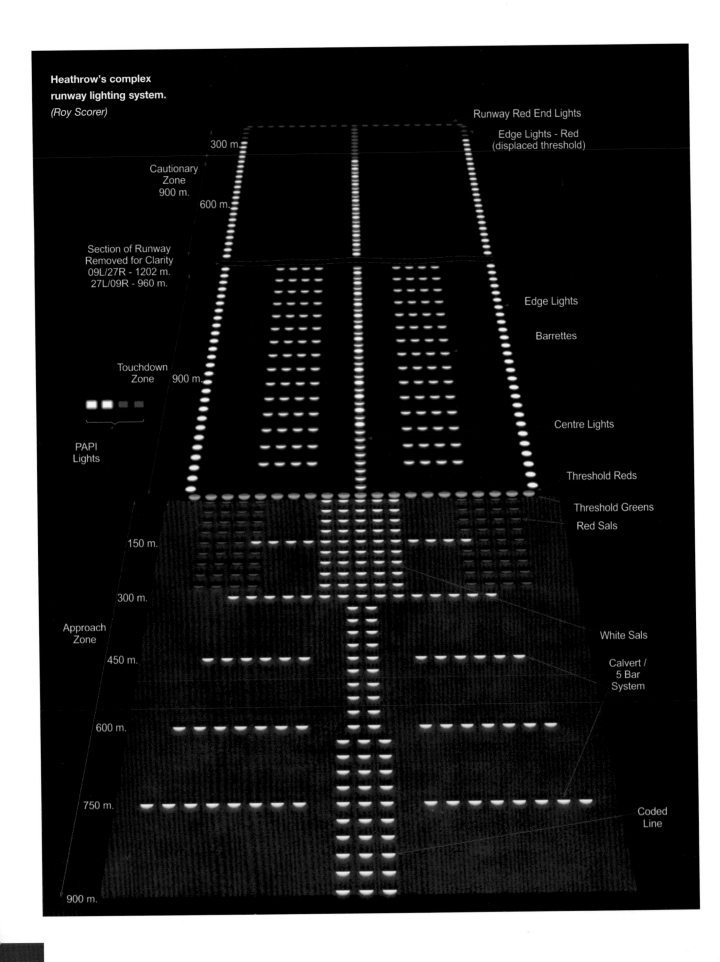

Heathrow's complex runway lighting system.

(Roy Scorer)

Runway Red End Lights

Edge Lights - Red
(displaced threshold)

300 m.

Cautionary Zone 900 m.

600 m.

Section of Runway Removed for Clarity
09L/27R - 1202 m.
27L/09R - 960 m.

Edge Lights

Barrettes

Touchdown Zone   900 m.

Centre Lights

PAPI Lights

Threshold Reds

Threshold Greens

Red Sals

150 m.

300 m.

Approach Zone

White Sals

450 m.

Calvert / 5 Bar System

600 m.

750 m.

Coded Line

900 m.

Lighting

There is a remarkable array of lighting provided at Heathrow in relation to the runways and taxiways. It is all controlled from the ATC tower and complies with ICAO standards. It includes:

Approach lighting

Approach lighting provides alignment, roll guidance and limited distance-to-go information for the visual completion of an instrument approach. The standard approach lighting system consists of a 900m coded line of white lights on the extended centreline of the runway, and five crossbars at 150m intervals. The bars decrease in width towards the runway threshold, and lines through the outer lights of the bars converge to meet the runway centreline 300m upwind from the threshold, coinciding with the aiming point. The following approach lights are installed at Heathrow:

■ High-intensity approach – approximately 900m, full Calvert-coded centrelines and five-bar system. The lights are unidirectional and white. Elevated approach lights are mounted on frangible gantries, posts or tripods. ('Frangible' means items of low mass, designed to break, distort or yield on impact, so as to present the minimum hazard to aircraft.)
■ Threshold lights – located at the threshold of the landing runway, consisting of a 50m bar. Lights are unidirectional, flush, high intensity and green.

ABOVE The supplementary approach lighting in use on runway 27L as an Emirates Airbus A380 touches down. *(Waldo van der Waal)*

LEFT Northern runway lights seen at night. *(HAL)*

■ Threshold wing bars – supplement the threshold lights and consist of four elevated, green, unidirectional lights either side of the threshold.

Supplementary approach lighting system (SALS)

Heathrow uses SALS to augment the approach lighting in low visibility operations. The system is operational on all four runways and consists of:

■ White SALS – two additional white lights on each side of the runway centreline, forming barrettes along the inner 300m of the approach centreline. The lights are unidirectional, high intensity and 1.2m apart.
■ Red SALS – red side row barrettes of four lights, spaced 1.5m apart, on each side of the white SALS barrettes, over the inner 270m of the approach lighting system. The lights are unidirectional and high intensity.

Precision approach path indicators (PAPI)

PAPIs form part of the approach lighting system at Heathrow and are used as a visual aid in conjunction with the ILS glide slope aerial to assist pilots in maintaining the correct visual approach. They are frangible and are installed adjacent to the runway, upwind of the threshold.

PAPIs consist of four units, each containing three high-intensity lamps. They are set to give a combination of red and white lights, indicating the correct angle of approach to the runway (they are set between 2.5° and 3.5°).

Runway lighting

Other than at night, runway lighting is used whenever visibility is less than 5km and/or the cloud base is less than 700ft. All aerodrome lighting is supplied from a ring main power supply. The ring main is backed up by standby diesel generators should there be power supply interruption.

Runway lighting forms an integral part of runway operations at Heathrow and primarily consists of:

■ Centreline lights – these are spaced at 15m intervals. The lights are bidirectional, flush, high intensity and beamed at 5° to the horizontal for the first 900m of each runway direction and 3° for the remainder. The lights are white to a point 900m from the runway end, with the following 600m alternate red and white, and the final 300m all red in colour to advise the pilot that he is nearing the end of the paved surface.
■ Edge lights – longitudinally spaced at 24.5m intervals and positioned 25m either side of the runway centreline. The lights are bidirectional and bielement, angled slightly towards the runway centreline, flush, high intensity and white, providing an indication of the edge of the paved surface.
■ Touchdown zone lights – longitudinally spaced at 60m intervals, covering the first 900m of runway available for landing. Consists of 15 rows of 4 barrettes each side of the runway centreline. The lights are unidirectional, flush, high intensity and white.
■ Runway end lights – a 50m row of lights located at the end point of the runway. The lights are unidirectional, flush, high intensity and red.

Taxiway lighting

The taxiway lighting system at Heathrow is very unique in that it is fully switchable, providing pilots with a traffic light system where they can 'follow the greens' and 'stop at the reds'.

The taxiway lighting system is so designed that in any particular section of taxiway only one route at a time can be selected, and all stop-bars are illuminated except those through which the selected route passes. This system, controlled by a Lighting Panel Operator in the ATC tower, provides aircraft with clear guidance on their taxiing direction and stopping locations.

The system consists of taxiway centreline

BELOW PAPIs are an essential visual aid used in conjunction with the ILS glide slope aerial. *(Roy Scorer)*

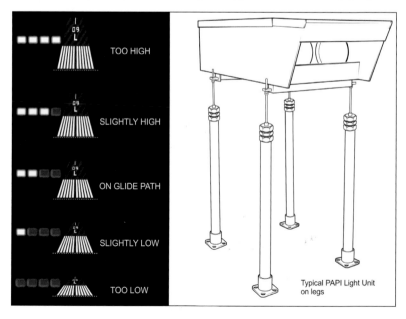

TOO HIGH

SLIGHTLY HIGH

ON GLIDE PATH

SLIGHTLY LOW

TOO LOW

Typical PAPI Light Unit on legs

lighting, taxiway stop-bars, illuminated signs and runway guard lights. The lighting has variable brilliance control (10%, 30% and 100%) depending on the conditions. At Heathrow the following taxiway lights are in use:

- Taxiway centreline lights – used for centreline guidance on taxiways.
- Runway turn-off/on lights – green and amber in colour, these are used for guidance to vacate or enter a runway.
- Red stop-bars – used to help protect runways against incursions, and to hold aircraft at a set location on the taxiway.
- Blue edge lights – used to denote the edge of a taxiway when centreline guidance is not available.
- Runway guard lights – commonly known as 'wig-wags' or 'guard ambers', these consist of two alternately flashing yellow lights and are situated at the taxiway/runway intersections to warn of approach to a promulgated runway.

Runway inspections

For Heathrow the areas of highest risk are arguably the runways, and ensuring safe and efficient runway operations requires regular inspections, which are carried out a minimum of four times a day by the ASD team in conjunction with ATC, as follows:

- At first light – ideally carried out before the main influx of arriving aircraft. During certain months of the year this inspection has to be carried out during the hours of darkness. The first-light inspection is a full, detailed surface inspection which entails several runs up and down the runway, and can take 30–45 minutes to complete. Close attention is always paid to any works that have taken place during the proceeding evening.
- Morning – ideally carried out mid-morning when traffic movements allow. This typically consists of a single run down the centreline of the runway using two teams at a speed that allows the inspection to spot any issues.
- Afternoon – as per the morning inspection.
- At last light – an inspection prior to night operations commencing.

These four visual inspections are twice the internationally recommended standards and have been carried out in this way at Heathrow since the

ABOVE **Runway inspection with the Airside Safety team.** *(Author)*

CHARLES DE GAULLE CONCORDE CRASH

The incident in 2000 at Charles de Gaulle when an Air France Concorde, Flight 4590, crashed showed the importance of runway inspections. The post-accident investigation found that five minutes before the Concorde took off a Continental Airlines DC-10 departing for Newark lost a titanium alloy strip measuring 435mm (17.1in) by 29–34mm (1.1–1.3in) during take-off from the same runway. During the Concorde's subsequent take-off run this piece of debris was lying on the runway and ruptured one of the Concorde's tyres. A piece of tyre debris then struck the underside of the aircraft's wing at high speed, sending out a pressure shockwave that ruptured the number five fuel tank at the weakest point, just above the undercarriage. Leaking fuel gushing out from the bottom of the wing was most likely ignited by an electric arc in the landing gear bay, or through contact with severed electrical cables. The aircraft crashed moments later, killing all 109 people on board and four people on the ground. A few days after the crash all Concordes were grounded. French authorities acknowledged that a required runway inspection was not completed after the Continental take-off, as was protocol for Concorde take-off preparation.

The main focus of the inspection is the general condition of the runway surface itself. However, crews are trained to assess the amount of rubber build-up, the condition of pits and drain covers, wear and clarity of illuminated runway signs and paint markings, runway lighting, patches of standing water and the state of any work in progress near the runways, as well as the condition and clarity of the airport's windsleeves. Debris found on Heathrow's runways to date has included a luggage strap, a suitcase lock, a bolt of unknown origin and even a fish dropped by a bird.

Additional runway surface inspections are carried out at the request of ATC, after a runway closure or after maintenance, immediately following an aborted take-off, after certain types of emergencies and always following a thunderstorm.

mid-1960s due to the high volume of traffic.

Maintenance

The runways at Heathrow represent some of the most intensively used transport infrastructure in the world, and as the surface has a finite life expectancy resurfacing is an essential maintenance task that comes round roughly once every 10–15 years. It represents a large-scale project for the airport and has to be carried out with the minimum amount of disruption and with a guarantee from the contractors to hand the runways back to the airport each morning to allow operations to resume.

In 2013 Heathrow resurfaced the southern runway at a cost of some £20 million, and in 2014 it will resurface the northern runway. Given the volume of traffic at the airport, this work is carried out at night and results in changes to the pattern of night-time flights as inbound and outbound aircraft are consolidated on to the one runway still in use.

Each evening after the final flight has landed the runway being resurfaced is closed and asphalt up to a depth of 50mm is removed across the width of the runway and along a length of 80m. An asphalt plant on site (located near Terminal 4) prepares material each afternoon for use that night. Over a 14-week period, more than 22,000 tonnes of old runway hard-wearing surface is removed and replaced.

A team of more than 150 operatives with more than 80 vehicles – including pavers, rollers, sweepers and a back-up for every item of plant on site – take responsibility for the resurfacing over a four-month period. Although use is being made of quick-drying tarmac, the work has to be scheduled to allow sufficient time for the newly laid surface to reduce in temperature for operational use. During each working night more than 1,000 cones mark out a temporary road system along the length of the runway to allow safe passage of vehicles and machinery. Every morning before the first flight of the day the team completes a thorough clean-up and hand-back procedure, with visual and physical sweeps of the runway.

Once the new asphalt has been laid, more than 80 miles of cabling and 1,000 new LED aeronautical ground lights (AGLs) are installed. AGLs help guide the pilots during poor visibility and are fitted within the surface of the runway. Work is also carried out constructing three new rapid exit taxiways (RETs) on the south runway and restoring all 13 entrances/exits on the northern runway. When the AGLs are fully installed, a flight check will be carried out (a small plane will conduct a series of landings on the runway) to make sure all the lights are operating correctly, and then the runway will be returned to its CAT III status, indicating that the runway can be used in all weather conditions.

Bird hazards

All birds on and in the vicinity of the airport are considered a threat to aircraft safety. Birds are small and fragile in comparison to aircraft, but a collision can have a shattering effect because of the high impact speed. The risk of a bird-strike is greatest at low altitudes, as that is where bird activity is concentrated, so Heathrow maintains a vigilant approach in this key area. The Airside Safety Department (see Chapter 7) has the responsibility for bird control and provides the operators, intelligence gathering and planning to carry out this service on a 24-hour basis. All staff involved in bird control undertake specialist firearms training to enable them to correctly identify the birds encountered at Heathrow, and all vehicles used are equipped with bird control equipment.

Research shows that 85% of bird-strikes occur at or near aerodromes during normal airfield operations (take-offs and landings), and a serious strike could lead to an aircraft accident with dire consequences.

Heathrow adopts a 'long grass policy', which recommends that the grass is kept to a length of 150–200mm (6–8in), since it has been found that this length deters certain birds species because it limits visibility, reduces populations of soil invertebrates and impedes access to the soil and ground surface, making prey more difficult to acquire.

There are various methods used by the team to deter birds, including:

- **Scarecrow Digiscare** – an electronic system, fitted inside the vehicle, with two forward-facing external loudspeakers on the roof. It is pre-programmed with the distress calls of various bird species. The general principle in broadcasting distress calls is to unnerve the birds and make them think one of their kind is in trouble. The intended reaction is for the birds to gain height and depart.
- **Arm scares** – by imitating the wing beats of a large raptor, most bird species will react immediately and fly away.
- **Culling** – when other dispersal techniques are ineffective, culling is used as a last resort.
- **Man presence** – people on foot are rare on airfields. A bird controller, simply by alighting from his vehicle and walking around, usually causes birds to become unsettled and depart.
- **Use of firearms** – pistols authorised for airfield use only fire either shell-cracker or blank

ABOVE The Griptester friction-testing machine. *(Findlay Irvine)*

with ice or snow.

The Griptester is a small three-wheel trailer that measures friction by the braked wheel, fixed slip principle. Its single measuring wheel, fitted with a special smooth-tread tyre, is mounted on an axle instrumented to measure both the horizontal drag force and the vertical load force. From these measurements the dynamic friction reading is automatically calculated and transmitted to a computer. This computer analyses and displays the data in a variety of ways. When one of Heathrow's runways is being tested, the screen automatically displays the average friction reading for each third of the runway as it is completed, and will also calculate and display the average friction readings for both sides of the runway together and over its full length.

cartridges to scare birds if necessary.

Griptester

During adverse weather conditions, Heathrow's Airside Safety Department (ASD) team makes use of a surface friction-testing machine called a Griptester to determine if there is sufficient grip for the runway to remain open. The device is used on runway, taxiway and apron surfaces that are wholly or partly contaminated

Rubber removal

When an aircraft lands its tyres are not spinning, and in the time it takes for them to get up to speed the tyres are effectively dragging on the runway as well as being put under pressure by the weight of the aircraft. The friction built up causes the rubber to polymerise and harden to the runway surface. This has a serious knock-on effect, as the rubber build-up affects the level of friction of the runway, most noticeably

RIGHT Heathrow's Trackjet vehicle is used for rubber removal from the runways. *(Author)*

as a reduction in braking and ground handling performance. This can lead to incidents such as runway overrun or a lateral slide off the runway, particularly during wet-weather movements.

The Griptester is again put to use for an initial assessment when runway rubber removal is scheduled. Once the information has been analysed, Heathrow's ASD team proactively tackle the issue of removing rubber from the runways by using a specially designed vehicle called a Trackjet, which uses high-pressure water, abrasives, chemicals and/or other mechanical means to remove the rubber build-up. The maintenance schedule for this work is generally based on the number of take-offs and landings, but the surface is also subject to regular visual inspections.

HEATHROW PEOPLE: MARK SANDFORD

Duty Manager Airside (DMA)

My role as DMA is to deliver the operational effectiveness of the aerodrome licence on a day-to-day basis and it's also my responsibility to ensure that normal service is resumed as quickly as possible after an incident.

To help make key decisions I use key air traffic and weather forecast information provided by the Heathrow Operational Efficiency Cell (HOEC), which is manned by NATS and the Met Office. Heathrow would normally land around 45 aircraft per hour, but this figure is dependent on many factors, such as aircraft type and weather conditions. By entering the wind conditions at the surface and at 3,000ft, Heathrow's landing rate prediction (LRP) tool provides an estimate of the amount of traffic that should be able to land.

If the demand drastically exceeds the LRP then a so-called 'flow rate' may need to be put in place to regulate the inbound traffic to an acceptable number. This rate has the potential to delay flights on the ground in Europe waiting to depart for Heathrow. The flow rate does not impact traffic from outside Europe, as these aircraft are more than likely already en route and have been for many hours.

During times of short-term delay we can approve the use of 'tactically enhanced arrivals measures' (TEAM), whereby a small number of aircraft are approved to land on the departure runway each hour to take some pressure off the inbound delay.

I chair four Heathrow operational conference calls (HOCC) each day with the airlines, NATS (Heathrow and Swanwick) and the Met Office. On the conference call we discuss all of the factors mentioned above, with the aim of 'tweaking' the operation where possible to obtain the most efficient outcome possible so that the high demand for capacity is managed at the highest level and in the most efficient way possible.

If aircraft are delayed into the night, it is my role to manage the airport's quota of night flights given to us by the DfT. There are strict rules on times, aircraft types and reasons for delay that we must abide by in deciding whether to approve a flight or not.

If any airside emergency should occur during a shift then I respond in whatever way is necessary to ensure that safety, efficiency and compliance are maintained throughout. The DMA represents Heathrow as a business at an incident and works closely with all emergency services at the scene.

CHAPTER 4

Air traffic control

OPPOSITE A controller observes the southern runway from the visual control room. *(HAL)*

Introduction

Air traffic control (ATC) is a highly complex operation at Heathrow when one considers that the airport has multiple terminals serving 86 airlines and 193 destinations. To put that into better perspective, using just two runways the 70 million passengers that pass through Heathrow each year touch down and take off from the airport courtesy of some 480,000 individual aircraft movements, each one of which is handled by a team of highly experienced air traffic controllers based in the airport's state-of-the-art tower in the centre of the airfield.

Heathrow's controllers are tasked with co-ordinating the movements of some 1,400 aircraft daily with a priority of keeping them at safe distances from each other, assisting in preventing collisions between aircraft and obstacles moving on the apron and the manoeuvring area, directing pilots around bad weather and ensuring that inbound and outbound traffic flows smoothly with minimal delays. A record of 1,389 movements was set on 1 September 2011 and official figures reveal that no two-runway airport in the world deals with this sort of volume of aircraft movements on a daily basis. However, compared to other airports Heathrow has very little 'wiggle room' – other airports with spare capacity can catch up on delays, but unfortunately

Heathrow does not have that luxury so minimising delays is absolutely vital. Even the smallest of hold-ups can cost a fortune and will invariably have a dramatic 'domino effect' across the entire airport, leading to frustrated passengers.

On approach to or outbound from Heathrow, flights must be guided by ATC through some of the most complex airspace anywhere in the world, exemplified by the fact that not only are there between 100 and 150 aircraft over London at any one time, but also that some 3,500 flights pass over southern England every day.

ATC services in the UK and at Heathrow are provided by National Air Traffic Services (NATS), whose controllers handle 22% of European air traffic but control only 5% of the airspace, with the UK acting as the 'gateway' for European flights to the United States. This chapter looks at the structure of UK airspace, the ATC personnel and the technology they use to deliver a round-the-clock service for Heathrow's flights.

Understanding UK airspace

Airspace over the UK is considered a 'national asset', and its safe and efficient utilisation is vital to both the UK economy and to national defence.

BELOW Airspace boundaries for the LACC and LTCC – a major confluence of airways, departures and arrival routes for the capital's five major airports. *(Roy Scorer)*

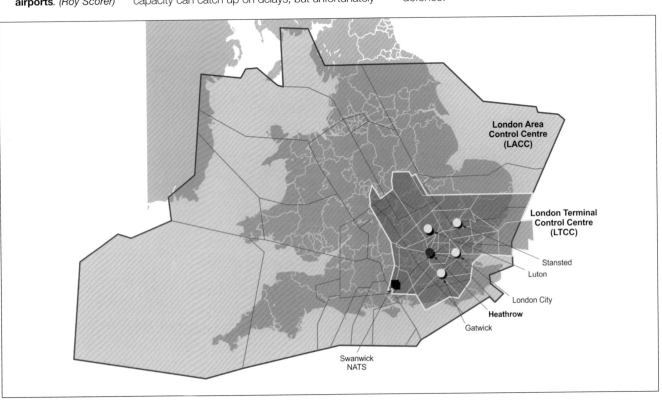

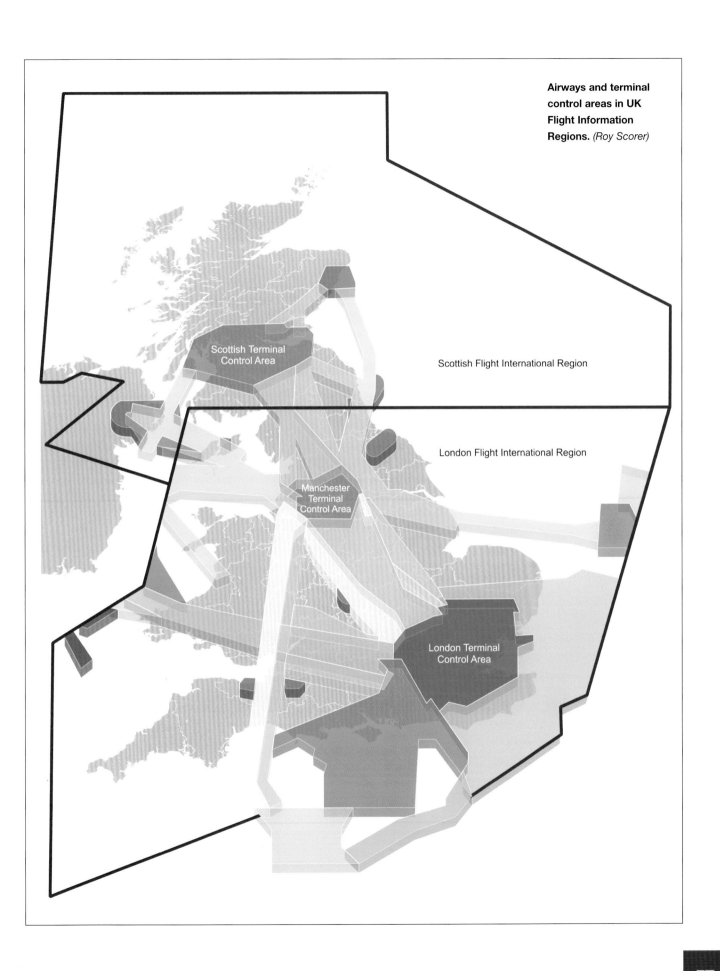

Airways and terminal control areas in UK Flight Information Regions. *(Roy Scorer)*

Scottish Terminal Control Area

Scottish Flight International Region

London Flight International Region

Manchester Terminal Control Area

London Terminal Control Area

To start with, UK airspace is divided into two Flight Information Regions (FIRs) – the 'London FIR' and the 'Scottish FIR', the boundary between them being the 55°N line of latitude. The London FIR comes under the London Area Air Traffic Control Centre and is controlled by NATS from Swanwick in Hampshire, while the Scottish FIR is under the Scottish Area Control Centre and operated by NATS from Prestwick.

Furthermore, the UK follows the International Civil Aviation Organization (ICAO) system, which classifies airspace so that pilots from anywhere in the world understand which flight rules apply and the nature of the ATC services on offer. The classification is based on the need to actively control access to airspace and the nature of the activity that takes place within it. The classification currently consists of seven categories as follows:

- Classes A, C and D require an air traffic control clearance to enter the airspace and receipt of an air traffic service is mandatory.
- Class G is uncontrolled in that any aircraft may use the airspace under the 'Rules of the Air', and although an air traffic service may be available it is not mandatory.
- Large portions of Class G airspace below 19,500ft are used extensively by military and general aviation.
- Classes E and F are not widely used in the UK and no UK airspace is currently designated as Class B.

From ground level to 66,000ft (20,100m), UK airspace is divided into two main categories – Controlled Airspace (CAS) and Uncontrolled Airspace (UCAS).

Controlled Airspace (CAS)

CAS is used to protect aircraft during the various phases of flight and to facilitate a safe and expeditious flow of air traffic. Any aircraft operating within CAS must have ATC clearance and be in receipt of an ATC service, ensuring that they observe any instructions issued. It is sometimes described as a 'known environment', as all traffic within the CAS is known to the ATC system. CAS requires a pilot to gain clearance from ATC and remain in two-way communication and obey the instructions of the controller. CAS is used almost exclusively by commercial passenger aircraft and is divided into five main types:

- Control Zones – these are established around major airports and extend from ground level up to a specified upper limit in order to provide protection for aircraft in the initial or final stages of a flight from or to an airport.
- Control Areas – these are located on top of the Control Zones and have a specified lower and upper level. They are mainly used when there are a number of busy airports located close together, providing protection for aircraft operating out of and into all of the airports.
- Terminal Control/Manoeuvring Areas – these are essentially a larger version of control areas and are best viewed as a confluence of airways and departure and arrival routes in and out of a number of airports that are located close together.
- Airways – these are corridors of controlled airspace, in the form of control areas, which provide the main routes connecting major airports.
- Upper Airspace – this comprises the majority of UK airspace from 24,500ft (7,467m) upwards.

Within CAS, standard routes are published as a template for planning purposes. However, controllers may use the full lateral and vertical extent of the CAS available to them. This is a vitally important part of their role because by tactically positioning aircraft within the CAS, controllers can ensure the most effective flow of traffic. As a result, routes do vary, and when there are fewer aircraft in the airspace – for example at night – it is possible for controllers to provide an aircraft with a routing that enables it to reach its destination in a shorter period of time than would be the case if the volume of air traffic dictated the use of a holding (queuing) system.

Uncontrolled Airspace (UCAS)

UCAS is airspace outside of CAS extending from ground level to 19,500ft (5,943m), or the base of CAS. It is airspace that is available to all civil users (including light aircraft, helicopters, hot air balloons and military aircraft). Any aircraft can operate in UCAS without talking to ATC and without specific ATC clearance by flying on a 'see and avoid' or

'see and be seen' basis, determining their routes based on their own requirements. There are air traffic control units that provide a Lower Airspace Radar Service (LARS) to aircraft that request a radar service, but the rules are entirely different to those within CAS.

Heathrow operations

Specifically in relation to operations at Heathrow, there are two key control centres at NATS in Swanwick:

■ **London Area Control Centre** (LACC) – which manages en route traffic in the London FIR, which can be upwards of 5,500 flights per day, making it the busiest Area Control Centre in Europe. This includes en route airspace over England and Wales up to the Scottish border.

■ **London Terminal Control Centre** (LTCC) – which handles traffic below 24,500ft (7,467m) flying to or from London's airports (Heathrow, Stansted, Gatwick, Luton and London City), as well as a considerable volume of overflying

BELOW **Inside the impressive LACC operations room at Swanwick.** (NATS)

traffic (for example, flights from Continental Europe headed to the United States – this airspace is controlled by LACC). Consequently the airspace for which the LTCC is responsible comprises large areas of CAS, the majority of which goes by the name of the London Terminal Control Area (LTMA) – a major confluence of airways, departure and arrival routes in and out of the region's major airports. This mixture of arrival and departure routes, climbing and descending aircraft, holding patterns and transit routes necessitates a myriad of highly developed, overlapping patterns and procedures that are closely controlled using extensive techniques. The LTMA can be likened to a massive motorway intersection, with routes crossing in all directions at multiple levels and with aircraft separated by a minimum of 1,000ft (305m) vertically or 3 miles (4.8km) laterally. The LTMA has a variable base level that starts at ground level in the control zones that surround the major airports – thus providing the maximum protection for aircraft in the most critical stages of flight. The base level steps up further away from the airports. This area, one of the busiest in Europe, extends south and east towards the coast, west towards Bristol and north to near Birmingham in the Midlands.

Heathrow itself has a Control Zone, which extends from the ground up to 24,500ft (7,467m). The TMA (Terminal Manoeuvring Area) sits above that and is classified as Class A airspace. The Heathrow radar controllers at Swanwick control the aircraft from the stacks and keep them within the confines of the Heathrow Radar Manoeuvring Area – a piece of airspace allocated to the Heathrow controllers that changes according to the direction of the landing runway. Its primary purpose is to provide an area specifically for the Heathrow controllers to vector aircraft from the stacks on to final approach.

Terminal Control (TC)

TC-based controllers at Swanwick provide air traffic services within the London Terminal Control Area (LTMA). TC is split into the 'area' and 'approach' functions.

The approach functions deal with the latter stages of flights inbound to Heathrow, Gatwick, London City, Stansted and Luton. Arrivals to the London airports are handed over from the LACC at Swanwick or the TC en route sectors, usually following Standard Terminal Arrival Routes (STAR) and are descended against the departing traffic, and sorted out into different levels, then routed to various holds, where they will hold until the

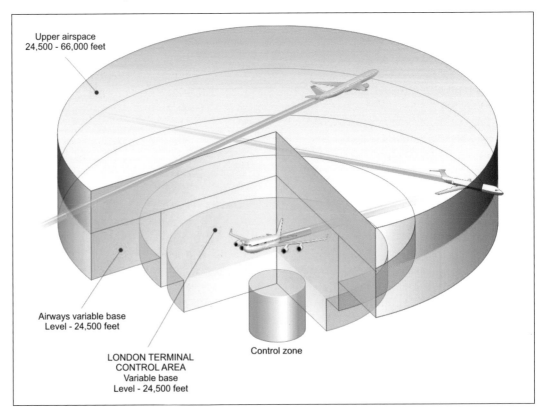

Upper airspace
24,500 - 66,000 feet

Airways variable base
Level - 24,500 feet

LONDON TERMINAL
CONTROL AREA
Variable base
Level - 24,500 feet

Control zone

RIGHT Dissecting UK airspace – some of the most complex anywhere in the world.
(Roy Scorer)

approach control units are ready to position them into an approach sequence to land. The majority of work for the approach units is controlling the sequence of aircraft from the holds until established on final approach about four miles away from the relevant airport.

The area function is based on two banks of controllers – TC North and TC South – which relate to the position of the airspace sector relative to Heathrow. Both TC North and TC South are further split into smaller sectors for control purposes. At its busiest, each sector is split and controlled by a radar controller. A sector co-ordinator is also present to co-ordinate with adjacent sectors or airfields and to release aircraft from the airfields. When the traffic is quieter, the sectors 'bandbox' – this groups different sectors together under the control of one radar controller. As an example, TC North East could include north-east departures, Lambourne (Heathrow arrivals and some Gatwick departures) and Lorel (Stansted and Gatwick arrivals). This would require three controllers.

Aircraft leaving Heathrow mostly depart on a free-flow principle – the radar controllers do not release each individual flight for departure, they just receive a pre-note via a computer system that the flight is pending. This cuts down on inter-unit co-ordination and allows the tower controller at the airport to decide the most efficient departure order. In many cases the aircraft's standard instrument departure (SID) routing does not conflict with the approach sequence of aircraft arriving at the airport, so the airport's approach control does not need to handle the aircraft and it is transferred straight to the LTMA controller on departure. The LTMA controllers then climb the departures through the arrivals (to the airports) that they are also working.

The role of NATS

Responsibility for the provision of ATC services in the UK lies with both civil and military service providers for aircraft within their area of responsibility. Traffic inside CAS is managed by NATS, who in turn are regulated by the CAA. The majority of NATS' activity is conducted from two ATC centres in Swanwick and Prestwick:

■ **Swanwick** – opened in 2002, Swanwick's state-of-the-art facility manages the flow of inbound and outbound traffic from the major

ABOVE **A controller at LACC Swanwick.** *(NATS)*

LEFT **The LACC radar display.** *(NATS)*

airports in the south-east of England, including Heathrow. The centre also manages the airspace south of 55°N (over England and Wales) in the Upper Airspace, along the Airways system and within the high levels of Control Areas. Controllers at Swanwick handle on average some 5,500 flights each and every day of the year. Swanwick alone controls 200,000 square miles of airspace above England and Wales, including the complex airspace of London. A small group of military controllers at Swanwick provide services to military aircraft operating outside controlled airspace. They work closely with civilian controllers to ensure safe co-ordination of traffic. The centre at Swanwick is the largest purpose-built air traffic centre in the world and is widely regarded as one of the most sophisticated IT projects in the country.

■ **Prestwick** – opened in 2010, Prestwick manages the flow of traffic in UK airspace north of 55°N (over Scotland) in the Upper Airspace, along the Airways system and within the high levels of Control Areas.

NATS is the UK's leading provider of air traffic services, maintaining the orderly, efficient and, above all, safe passage of aircraft through UK airspace and beyond. It controls 2.5 million square miles of airspace above the UK and eastern North Atlantic and handles more than 2 million flights per year. Of these flights, around 87% are from or to the UK. The remaining 13% are over-flights.

NATS provides services worldwide 24 hours a day, 365 days a year at 15 major commercial airports in the UK and Gibraltar, under competitive contract to different airport operators. The business also consults on and delivers new solutions to airports to improve their ATC operations.

NATS has a key role to play in reducing aviation's CO_2 footprint. By improving airspace design and procedures and working closely with airports and airlines, it can help aircraft achieve optimal flight profiles. This includes better flight planning, better altitude and speed profiles and shorter, more direct routes.

NATS' en-route charges are paid directly from a route-charging structure that is based upon the size of an aircraft and the time/distance spent in UK airspace. NATS services at Heathrow are paid for by the airport, which in turn recovers the cost from airlines at a rate per landing capped by the CAA.

Heathrow's holding stacks

If traffic is not particularly heavy, inbound aircraft are vectored directly to intercept the ILS for the landing runway at Heathrow, but given the volume of flights landing at the airport it is not uncommon on an inbound flight – especially during early morning arrivals when the airport is at its busiest – for passengers to hear the captain announce that 'It's a busy morning at the airport and we've been asked by air traffic control to take a position in the hold.'

When the airport is busy, aircraft orbit in a holding pattern. With around 700 aircraft arriving every day, the holding stacks effectively buy air traffic controllers some time to better sequence inbound flights.

The landing sequence for an inbound aircraft generally starts by being directed to one of four main reporting points ('stacks' or 'holds'). These are Bovingdon (BNN) over Hertfordshire (north-west of London), Lambourne (LAM) over Essex (north-east of London), Biggin Hill (BIG) over Bromley (south-east of London) and Ockham (OCK) over Surrey (south-west of London).

THE LONDON TERMINAL CONTROL CENTRE (LTCC)

The London Terminal Control Centre was an air traffic control centre based in West Drayton just north of Heathrow. Operated by National Air Traffic Services (NATS), it provided air traffic control services to aircraft arriving and departing from six London airports and one Royal Air Force station, plus en route services to other aircraft that entered its airspace.

The centre was originally opened as RAF West Drayton, a military air traffic control facility. The civilian control function subsequently became the London Air Traffic Control Centre (LATCC), operating alongside the RAF. In the early 1990s the 'Central Control Facility' (CCF) was formed within the centre to provide terminal control services to aircraft arriving at and departing from the main London airports.

In 1992 the Heathrow and Gatwick approach control units moved to West Drayton to share facilities with the CCF. The CCF and the two approach control units were moved into the new Terminal Control room in 1995, and thus became a separate entity to the London Area Control Centre. To reflect the fact that there were now two civil control rooms (Area and Terminal) the centre was renamed the London Area and Terminal Control Centre, while retaining the same LATCC abbreviation.

RAF West Drayton formally closed in the 1990s, though military personnel remained on site until 2008.

Civilian operations at the centre ceased in November 2007, after Terminal Control moved to Swanwick to be reunited with the LACC. Military operations moved to a new control room, also at Swanwick, in January 2008.

BELOW Heathrow's four holding stacks. *(Roy Scorer)*

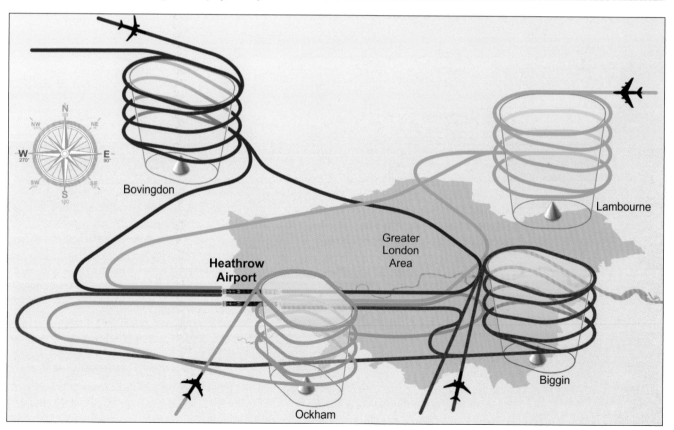

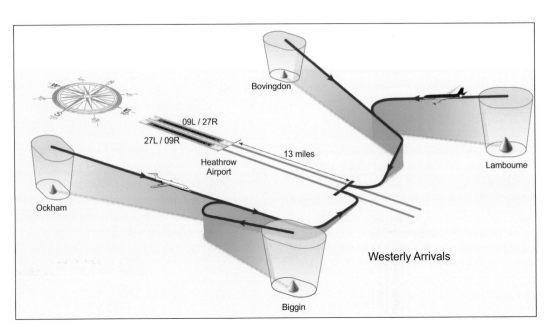

RIGHT Heathrow's Westerly and Easterly arrivals patterns – aircraft from the four stacks are 'knitted together' by controllers working in close co-operation. Aircraft are then established on the ILS for their final approach. *(Roy Scorer)*

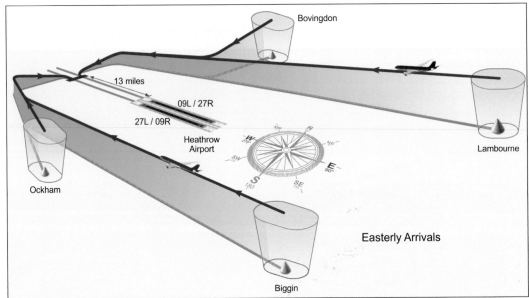

Each of the stacks is defined by a VHF omnidirectional radio range (VOR), a type of short-range radio navigation system used as a standard all over the world, enabling aircraft to determine their position and stay on course by receiving radio signals transmitted by a network of fixed ground radio beacons.

The stack is an area of airspace with a track distance anywhere from 8 to 20nm in size. New planes enter the stack at the top and are directed to move down as others exit below them. The stack effectively acts as a 'waiting room', allowing controllers to efficiently organise planes for landing.

Within the stacks, controllers keep aircraft 1,000 vertical feet (305m) apart and between 7,000ft and 15,000ft above the ground. The number of stacks in use and the number of aircraft in each stack depends on how busy the airport is at the time and the direction from which the flights are coming. If these holds become full, aircraft are held at more distant points before being cleared onward to one of the four main holds. The stacks have been in the same locations since the 1960s.

NATS air traffic controllers for Heathrow Approach Control (based at Swanwick) then guide the aircraft to their final approach, merging aircraft from the four holds into a single stream of traffic, sometimes as close as three miles apart. When Heathrow's ATC is ready for an aircraft to land, the pilot is instructed to move out of the holding pattern and begin his descent to the airport. The descent is usually begun at approximately 22nm from touchdown from 7,000ft. This allows the aircraft to fly making minimal power setting changes. In turn, this reduces fuel burn and minimises noise pollution. This is known as a continuous descent approach (CDA). Once the aircraft is established on its final approach, control is handed over to Heathrow Tower for the final part of the flight.

There is no set route between the stack and the start of the final approach because there are many variables for air traffic control to consider, such as weather and the position of other aircraft. By the time the aircraft is within 7.5nm (13.9km) of touching down it must have lined up with the airport's Instrument Landing System.

Heathrow's air traffic controllers

The full ATC service at Heathrow is provided by staff from NATS, who have the challenging task of dealing with the 480,000 air traffic movements at the airport each year. A pool of 65 air traffic controllers (ATCOs) work across five watches (shifts) in the tower and are trained in all aspects of ATC operations, allowing them to fulfil all of the key ATCO roles in the tower.

An ATCO's roster is typically based on working six shifts followed by four days off. Even though there are night noise restrictions in place at Heathrow, which limit the number of flights and a quota on the amount of noise permitted, the airport is open 24 hours, and hence a 24-hour ATC service is required every single day of the year. As a result shifts run from 07:00 to 14:30, 14:30 to 22:00 and a night session from 22:00 to 07:00. Generally speaking there are no scheduled flights between 23:30 and 04:30; however, occasionally aircraft operate after 23:30 due to delays either here or in other airports around the world.

From their unique vantage point in the Visual Control Room (VCR), Heathrow's ATCO team operate from what many consider to be the nerve centre of the airport. Sitting calmly above the airport, some of the world's best controllers manage a fluid jigsaw, orchestrating the landings, take-offs and ground movements of hundreds of millions of pounds worth of aircraft every day. The team are surrounded by computers and state-of-the-art technology, but at the end of the

day all core decision-making is done by humans, and being up in the VCR gives the team a clear overview of everything going on at the airport.

Heathrow's ATCOs are different to those at NATS in Swanwick, who work purely off radar information. As 'visual controllers' Heathrow's ATCOs not only rely on what the screen in front of them is saying, but more importantly on what they can physically see when looking out of the VCR windows, so much so that the two key positions of Air Arrivals and Air Departures are physically elevated above the other positions in the VCR to afford the best possible view.

The Heathrow VCR typically has the following 14 personnel on station:

ABOVE Air traffic control demands unobstructed views of the airport and its approaches. The tower provides a clear 360-degree cone of vision using tapered glass panels. *(NATS)*

Supervisor	1	Sometimes referred to as the 'Watch Manager', he is responsible for overseeing the entire VCR operation, making tactical decisions around air traffic and flow, and liaising with all other agencies that interact with ATC. He also ensures the ATCO team adhere to their regulated shifts, which require working periods of no more than 90 minutes at a time, with 30-minute breaks in between.
Ground movement planner (GMP)	1	Regulates the start-up of aircraft to ensure that the ground movement controllers are not overloaded, that aircraft don't unnecessarily waste fuel by holding before take-off, and generally work to ensure an expeditious mix of traffic heading to the departure runway.
Ground movement controllers (GMC)	3	These controllers are responsible for all aircraft vacating the runway and taxiing to gate or taxiing from gate to holding point, all aircraft pushing back from stand, and all towing movements. The complexity of movements on the ground at Heathrow is just about as complex as those in the air, so the task is split into three key areas: GMC1 controls the northern half of the central terminal area, GMC2 controls the southern half of the central terminal area, and GMC3 controls the area around T5.
Lighting panel operators (LPO)	3	The LPOs operate Heathrow's unique and complex lighting system on instructions from the GMC controllers, and use a panel that shows a scale map of the airfield. They use switches on the panel to set a lighting path for aircraft to follow in the corresponding section of the airport. They listen in on the appropriate GMC frequency and set the path to the gate as directed by the GMC controller. At night and in low visibility the lights effectively provide a route protected from other aircraft, especially when vacating runways.
Air Arrivals Controller	1	Responsible for the arrival runway. This controller ensures the separation between the landing aircraft is not eroded (and positively intervenes if it is) and gives safe landing clearances to aircraft. He also ensures that the runway is available for the next inbound aircraft, as well as integrating crossing vehicular traffic into the pattern. With a conveyor belt of inbound traffic, the arrivals controller tries to maximise the use of the runway at all times.
Air Departures Controller	1	Responsible for the departure runway. Ensures the most expeditious throughput of traffic is achieved, maximising the departure separations, taking into account vortex wake, slot times and routings. Also integrates landing aircraft when there are delay situations, and vehicular traffic also.
Support position or air traffic service assistant (ASTA)	1	Assists the controllers with airline liaison, agencies, weather-watching and general observation needs.
Trainee ATCOs	2	Heathrow needs to continuously train and validate staff to keep up its operational requirement. Trainees must complete a six-month course at the College of Air Traffic Control before joining Heathrow Tower, where they undergo specific on-the-job training with valid ATCOs.

Elsewhere in the tower are the NATS Operations, Training, Administration and Engineering departments, which support ATC activity at Heathrow.

ABOVE AND BELOW Key positions in Heathrow's visual control room. *(Author)*

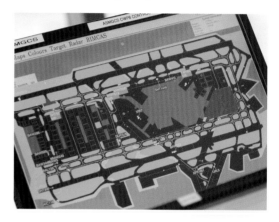

One of the biggest challenges for Heathrow's ATCOs comes from fog, which impacts on visibility and requires increased separation between aircraft. If there is more space between planes, not as many planes can land or take off each hour – which can mean delays and cancellations.

A 'normal' landing rate at Heathrow is around 45 aircraft landing per hour. With reduced visibility the arrival rate drops to around 36, and under low-visibility procedures (LVPs) the rate drops to around 24 aircraft landing per hour.

Fog has major implications for Heathrow's ground controllers – although aircraft are equipped with technology that enables them to fly through

WHAT DOES IT TAKE TO BE AN AIR TRAFFIC CONTROLLER?

It is often said that air traffic controllers must be able to think and visualise in three dimensions. An air traffic controller is a specialised role and individuals must be able to gather information from what they hear and see, then use that information to make decisions quickly to be able to move aircraft safely through their area of responsibility. They must be able to read and interpret data as well as predict the whereabouts of aircraft from headings and speeds, and they must be able to concentrate intensely for the period of time they are on position.

Students train at the College of Air Traffic Control, where they do a basic course and are then streamed on to a particular rating discipline – Area, Approach or Aerodrome. On each course there is a mix of theoretical learning and practical simulator runs that recreate real air traffic situations for practical training. There are regular exams to pass during the courses and a student must pass all of these to be successful. The instructors have been controllers themselves.

After graduating from the college, students will be posted to a unit – either an Area Control Centre or an airport control tower – and continue to train. They need to learn the geography and procedures of the sectors/tower they will control and apply the skills learnt at the college to them. Once they are ready to work on their own they will sit a final exam, a validation board. The entire process from starting at the college to validation takes, on average, three years.

Given the nature and intensity of the role, controllers are subject to continuous assessment for as long as they continue to control.

BELOW The visual control room in operation at night, looking west towards Terminal 5. (HAL)

fog, the signal that aircraft follow from the Instrument Landing System (ILS) must be protected to a greater distance from the runway. In low-visibility procedures the preceding aircraft has to be allowed to land and clear the Sensitive Area of the ILS before the following aircraft is given landing clearance. When LVPs are in use at Heathrow the spacing on approach is increased from the minimum of three miles to six miles to ensure that aircraft have cleared the Localiser Sensitive Area before the following aircraft is given landing clearance.

When severe disruption at Heathrow is expected, a decision may be made to reduce the flight schedule in advance. This decision is made by a group comprising representatives from Heathrow, NATS and the airlines, who collectively agree how much the schedule must be reduced by in order to cope with the prevailing conditions. If Heathrow operated at lower-capacity levels this sort of action would not be necessary, or, if it was

required, it would be on a far smaller scale than is currently the case, with the delays effectively absorbed into the reserve capacity.

ATC operations

Arrivals

Once an inbound aircraft nears the stack it is released from TC to Heathrow Approach (in Swanwick). The pilot is either told to enter the hold or is vectored into the landing sequence.

The Heathrow Intermediate Director North and Heathrow Intermediate Director South work closely together to create two orderly streams of inbound aircraft from the north and the south towards the final approach path by giving instructions to aircraft on their required height, speed and heading. The Final Director then works to 'knit' the two streams into a single line of approaching aircraft, ensuring they are sequenced with the correct amount of separation (the 'landing interval').

Once the 'knitting' is done and the two streams are merged, aircraft are established on the ILS (anywhere from six to eight miles out from Heathrow) and control passes to the Air Arrivals Controller in the tower. At this point, separation is closely monitored until a safe landing can be passed. The flight crew then receive their landing clearance and details of the local weather conditions at the airfield.

The direction of the wind is assessed at the airport at ground level and at 1,000ft and 2,000ft by ATC, and also with reference to reports from aircrew.

Although the direction of the wind is the predominant factor that affects which direction aircraft must land (and take off in) at Heathrow,

government policy is that unless the wind is too strong planes should always take off to the west (towards Windsor). This means that unless the wind is too strong, they should therefore arrive from the east (over London). This is known as the 'westerly preference'. However, if the surface wind is from the east and is over 5kt, aircraft will take off to the east and so arrive from the west (over Windsor).

A system of runway alternation – introduced in the 1970s – for aircraft landing during westerly operations (ie when aircraft arrive over London) is possible because Heathrow has two runways that are parallel to each other, so aircraft can arrive from the east but land on either the northern or southern runway. The reason for landing runway alternation is to provide predictable periods of relief from the noise of landing aircraft for communities under the final approach tracks to the east of the airport. Runway alternation applies to landing aircraft, hence aircraft taking off during westerly operations can use either runway, but most use the runway that is not in use for arrivals.

The present pattern provides for one runway to be used by landing aircraft from 06:00 until 15:00 local time. The other runway is then used from 15:00 local until after the last departure.

On Sunday each week the runway used before midnight continues to be used for landings until 06:00 local. This means early morning arrivals before 06:00 local use a different runway on successive weeks and that the runways used by landing aircraft before and after 15:00 local also alternate on a weekly basis.

In some circumstances ATC may use both runways for arrivals for a short period. This is referred to as 'tactically enhanced arrivals measures' (TEAM), and helps to increase the overall landing rate and reduce inbound delays. This technique requires the ATCO team to precisely interweave aircraft, and has been shown to be most beneficial during periods of high demand, such as early morning arrivals. Once an aircraft has landed and vacated the runway it is transferred to the ground movement controller, who provides directions to the parking stand.

Departures

When an aircraft is ready to depart from Heathrow, the pilot calls the GMP for permission to push back and start the aircraft's engines. With this under way he contacts the GMC, who determines when it is safe to depart the gate. The GMC takes into consideration a host of factors, including the volume of traffic on the airfield and how many other aircraft have already started up, in order to minimise ground delays and save fuel.

When satisfied with the conditions, the GMC issues the push-back clearance from the stand together with directions to the departure runway. As the aircraft approaches the runway, control is passed to the Air Departures Controller, who puts the aircraft into a departure sequence designed to ensure the most efficient use of the runway. It is not uncommon to see a number of different aircraft being held in the holding area as the take-off sequence is shuffled into the most efficient departure order.

As a rule of thumb, ATCOs alternate departures between straight ahead, left and right. Two identical aircraft taking off in succession and heading in different directions would typically be permitted to depart one minute apart, as their headings do not conflict and there are no significant wake turbulence issues. Unfortunately for the ATCOs at Heathrow, there is a big mixture of aircraft that use the airport, and more than 190 possible destinations, so departure procedures are a complex affair. With significant wake turbulence issues to consider and conflicting headings, the departure interval may need to be increased.

All the while, the GMC keeps a close eye on all of the airport's taxiways, both visually out of the VCR's windows and by using the ground radar to track aircraft, ensuring they take the correct routes either to their destination stand or to the runway, and do not come into conflict with other manoeuvring traffic.

Once line-up clearance has been granted, pilots are expected to taxi on to the runway as soon as the preceding aircraft has started rolling down it, which again helps to improve runway efficiency.

Outbound aircraft are normally spaced either

BELOW Controllers at Heathrow sequence aircraft of different sizes in a particular way prior to departure so as to maximise the efficiency of the runways. *(Author)*

Air traffic control **69**

be consistently alternated, anything between 40
and 55 aircraft can depart in an hour.

Aircraft follow flight paths known as 'noise
preferential routes' (NPRs) up to an altitude of
4,000ft. NPRs were set by the DfT in the 1960s and
were designed to avoid aircraft flying over built-up
areas where possible. They lead from the take-off
runway to the main UK air traffic routes, and form
the first part of the SIDs. The routes have not been
altered since they were established, in order to give
people the predictability of knowing where noise
from departing planes will be heard. Their location
remains the responsibility of the government and the
airport has no authority to change them.

Aircraft cannot fly in the same way that a train
runs on tracks. This means that there will be some
variation as to where different aircraft will be on the
NPR. This is because all aircraft perform differently,
and speed, wind, weight and temperature affect
the performance of an aircraft and can cause
it to drift left or right. It is for these reasons that
each NPR has a 'swathe' measuring 1.5km either
side of the route centreline, resulting in a corridor
3km wide. As long as the aircraft are within this

one to two minutes apart depending on their
routing (north or south) and vortex wake category.
Aircraft depart using Standard Instrument
Departure (SID) routes designed to affect the
lowest number of people with the least amount of
noise possible. If north and south departures can

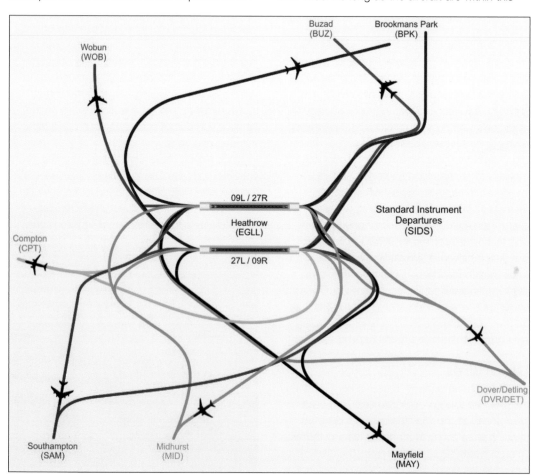

'swathe' they are considered to be on track. They do not have to follow the centreline of the NPR.

Once aircraft have reached 4,000ft, ATC can instruct the pilots to leave the NPR and fly a more direct heading to their destination. This is known as vectoring.

Understanding wake turbulence

One of the single biggest factors which influences capacity levels at Heathrow and the speed at which inbound aircraft can land and outbound aircraft can take off is the phenomenon of 'wake turbulence'.

As an aircraft passes through the air, it creates turbulence. This disturbed air principally comes from two key areas: firstly, from the engines, which expel rapidly moving gases – this air is extremely turbulent but short in duration, and is sometimes referred to as 'jet wash'; and secondly, from the effect of the rotating air masses generated behind the wingtips of an aircraft. In contrast to jet wash, this air is much more stable and can be encountered for up to three minutes after the passage of an aircraft.

Wingtip vortices occur when a wing is generating lift. Air from below the wing is drawn around the wingtip into the region above the wing by the lower pressure there, causing a vortex to trail from each wingtip. The strength of wingtip vortices is determined primarily by the weight and airspeed of the aircraft – the heavier the aircraft and the more slowly it is flying, the stronger the vortex. As a result this has a direct bearing on capacity at Heathrow and is a key factor for the ATC team to consider when sequencing aircraft for both take-off and landing.

Wake turbulence is especially hazardous in the region behind an aircraft in the take-off or landing phases of flight. During these phases, aircraft operate at high angles of attack and this flight attitude maximises the formation of strong vortices. In the vicinity of an airport there can be multiple aircraft, all operating at low speed and low height, and this provides extra risk of wake turbulence, with reduced height in which to recover from any upset.

To avoid dangerous wake vortex encounters, official weight-dependent separation distances have been established for approach and landing. The strength of the vortex is governed by the weight, speed, and shape of the wing of the generating aircraft. The vortex characteristics of any given aircraft can also be changed by the extension of flaps or other wing configuration devices, as well as by a change in speed. The greatest vortex strength occurs when the leading aircraft is heavy, clean, and slow. For this reason, as the table shows, the separation minima for a light aircraft landing behind an A380 is 8nm. To put this into further perspective, even if the following aircraft is another A380, a separation minima of 4nm is required.

Heathrow's ATCOs sequence aircraft landing and taking off using the following minima:

Landing		
Leading aircraft	Following aircraft	Wake turbulence separation minima distance (nm)
A380–800	A380–800	4
A380–800	Heavy	6
A380–800	Upper and lower medium	7
A380–800	Small	7
A380–800	Light	8
Heavy	A380–800	4
Heavy	Heavy	4
Heavy	Upper and lower medium	5
Heavy	Small	6
Heavy	Light	7
Upper medium	A380–800	#
Upper medium	Heavy	#
Upper medium	Upper medium	3
Upper medium	Lower medium	4
Upper medium	Small	4
Upper medium	Light	6
Lower medium	A380–800	#
Lower medium	Heavy	#
Lower medium	Upper and lower medium	#
Lower medium	Small	3
Lower medium	Light	5
Small	A380–800	#
Small	Heavy	#
Small	Upper and lower medium	#
Small	Small	3
Small	Light	4
Light	A380–800	#
Light	Heavy	#
Light	Upper and lower medium	#
Light	Small	#
Light	Light	#
# signifies that separation for wake turbulence reasons alone is not necessary.		
Source: Civil Aviation Authority (CAA).		

Take-off		
Leading aircraft	**Following aircraft**	**Minimum wake turbulence separation at the time aircraft are airborne**
		Departing from the same position or from a parallel runway separated by less than 760m
A380–800	Heavy (including A380–800)	2 minutes
	Medium (upper and lower)	3 minutes
	Small	3 minutes
	Light	3 minutes
Heavy	Medium (upper and lower)	2 minutes
	Small	2 minutes
	Light	2 minutes
Upper and lower medium or small	Light	2 minutes
Leading aircraft	**Following aircraft**	**Minimum wake turbulence separation at the time aircraft are airborne**
		Departing from an intermediate point on the same runway or from an intermediate point of a parallel runway separated by less than 760m
A380–800	Heavy (including A380–800)	3 minutes
	Medium (upper and lower)	4 minutes
	Small	4 minutes
	Light	4 minutes
Heavy (full-length take-off)	Medium (upper and lower)	3 minutes
	Small	3 minutes
	Light	3 minutes
Medium or small (full-length take-off)	Light	3 minutes

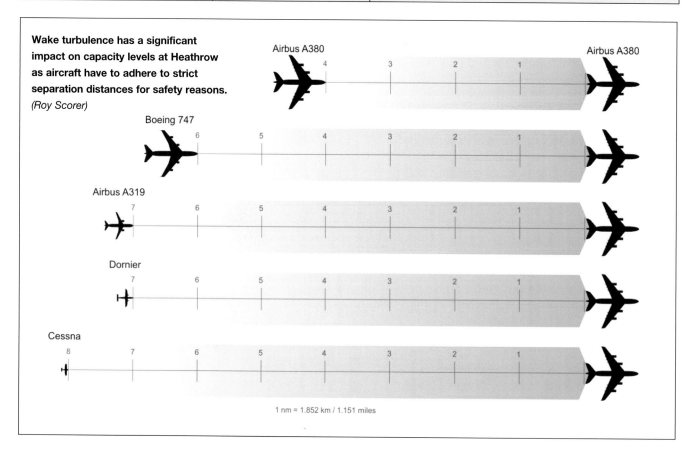

Wake turbulence has a significant impact on capacity levels at Heathrow as aircraft have to adhere to strict separation distances for safety reasons. *(Roy Scorer)*

Airbus A380

Boeing 747

Airbus A319

Dornier

Cessna

1 nm = 1.852 km / 1.151 miles

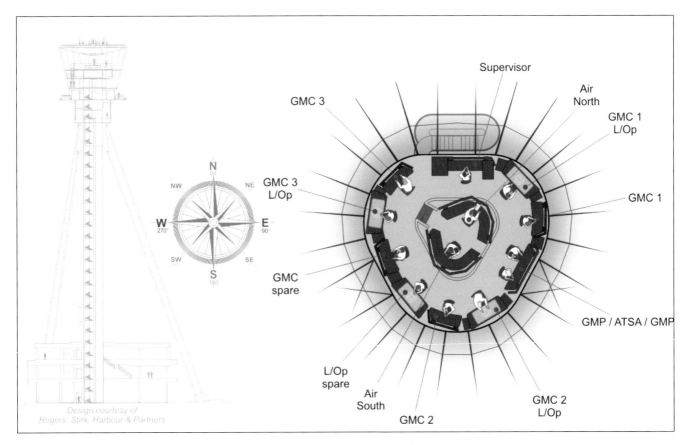

Note: ATC shall apply the minima as prescribed overleaf, irrespective of any pilot request for reduced wake turbulence separation. ATC does not have the discretion to reduce wake turbulence separation minima.

Source: Civil Aviation Authority (CAA).

If two or more aircraft experience a loss of minimum vertical or horizontal separation, a conflict is flagged up. This does not in itself suggest that the aircraft are at any risk of collision but it is a controller's job to prevent this situation from occurring.

Separation distances are under regular review in conjunction with the CAA to discover how possible margins might be used to improve current and future capacity issues.

The ATC tower structure

Heathrow's air-traffic controllers have to maintain constant visual contact with aircraft on the ground, but the development of T5 introduced obstructions to the required sightlines between the existing tower and aircraft using the new terminal, so a new and taller tower at a different location was needed. The optimum tower dimensions were calculated by assessing the sightlines to all taxiways and stands, while the best location was determined as the airport's geographic centre.

The new tower was opened in April 2007 and is one of the defining features of the airport. At

87m (285ft) tall, twice the height of the original tower, it affords controllers an unobstructed 360° view of the entire airfield. It is the tallest tower in the UK – twice the height of Nelson's column – and plays host to 150 staff.

Given its location in the centre of the airport, construction had to be engineered so as not to interfere with the day-to-day running of the airport. In essence, the tower was constructed on an 'island site' surrounded by aircraft, with its central steel column progressively raised one section at a time over several nights, while no planes were flying.

The operation to move the tower's cone-shaped top section into place was carried out slowly on three remote-controlled hydraulic 144-wheel flatbed trucks, along taxiways and across the southern runway, by a specialised team, who had to take great care to spread the huge load and not cause any damage to the airfield – the 900-tonne load greatly exceeded the 400 tonnes of a fully loaded Boeing 747 for which the pavement was designed. Damage to the runway or breakdown of the transporter en route would cause effective closure of the airport, with resultant damages likely to exceed half the value of the entire control tower project, valued at some £50 million.

After a 24-hour delay due to thunderstorms, the overnight move was achieved without incident in less than two hours amidst a sea of press and TV cameras. At the control tower site, the cab was manoeuvred and placed on to its foundation to within 10mm of dead centre. The final design provides what is thought to be the largest cone of vision of any control tower in the world.

Extensive wind-tunnel modelling was undertaken to optimise the tower's aerodynamic performance by reducing the drag and crosswind response of the design. Small aerodynamic strakes (stabilisers) are attached to the side of the mast to control vortex-shedding and significantly reduce the cross-wind response.

The entire structure is anchored to the ground by three pairs of cable stays. The mast provides access to the visual control room via one internal and one external lift, plus an enclosed stairway of 444 steps. The tubular plated main core was shaped to accommodate lifts, stairs, services and electronics, and the three stays ensure the rigidity criteria for the radar systems.

A three-storey building at the base of the tower contains the NATS offices, administration and training rooms, technical equipment areas, and main plant rooms. The tower is accessed via a 100m pedestrian bridge link from Terminal 3's Pier 7.

Information provided courtesy of *The ARUP Journal.*

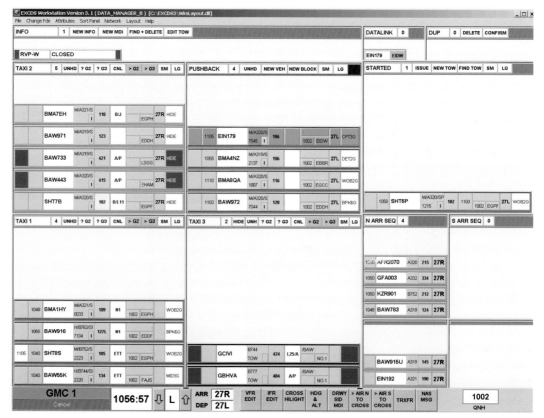

RIGHT Electronic flight progress strips allow for simplified handling of flight data and reduce a controller's manual workload significantly. *(NATS)*

Technology

Heathrow's ATC team rely on a remarkable array of technology when it comes to daily operations, including:

Electronic flight progress strips (EFPS)

Whether in electronic or paper form, flight progress strips have played an integral part in ATC operations since their introduction in 2007. As one of the world's most technologically advanced airports, Heathrow makes use of EFPS to display flight data as the primary visual aid used by ATCOs to separate and sequence aircraft. Each EFPS carries all the information an ATCO needs to know about a flight. As instructions and information are given to the flight crew, these are also recorded on the EFPS.

Different coloured strips are used to identify whether flights are arriving, departing or transiting. The information on an EFPS can be customised to provide more or less information to ATCOs, but typically includes:

- Aircraft identification (*eg* aircraft registration or a flight number).
- Aircraft type using the relevant four-letter ICAO designator (*eg* B744 for a Boeing 747-400).
- Level (assigned altitude).
- Departure and destination.
- Wake vortex category.
- Stand information.
- SID information on departures.
- CTOT (slot times).

The EFPS system can be integrated with other tower and airport systems and can be customised to support the specific needs of the controllers, making it far more effective than the paper equivalent. Traditional paper strips have several limitations: they are time-consuming to print and update, the information on the strips stays with the controller, and the possibilities for integration with safety nets are limited.

It is interesting to note that the term 'handover', which is used today to denote the computerised transfer of control of an aircraft from one sector or controller to another, comes from the older technique of physically handing over the flight progress strip to the next controller to denote the transfer of responsibility. This technique is still used today within some control towers; for example, the ground controller may physically hand the strip to the local controller as the aircraft reaches the runway, or the local controller will drop the strip down a chute to the departure radar controller in the room below, once the aircraft has been cleared for take-off.

Advanced Surface Movement Guidance Control System (A-SMGCS)

A-SMGCS is a surface movement radar that interrogates the secondary radar of aircraft and produces a data-block on the screen for each aircraft. This shows the call sign of the aircraft as well as whether the aircraft is an inbound or an outbound. This allows for improved situational awareness and also gives much better performance in low-visibility conditions. The system also allows for clearances to be passed to aircraft by ground movement controllers based solely on radar data, so the aircraft involved do not necessarily have to be visible from the VCR.

Surface movement radar (SMR)

The basic radars that show primary radar returns of vehicles and aircraft on the manoeuvring area of an airport. Used by ground movement controllers to help their awareness of where aircraft are.

Aerodrome traffic monitor (ATM)

A radar display of the area around an aerodrome. Used by ATCOs to identify aircraft, establish initial departure separations and to monitor arrival separations to ensure that they do not erode below the permitted minima.

ABOVE Surface movement radar is placed in key positions across the airfield to aid ground movement controllers. *(Author)*

RIGHT Heathrow benefits from a fully switchable ground lighting system, the only one of its kind in the world. *(HAL)*

Airport display information system (ADIS)

Used to display all of the most recent meteorological information that controllers need and have to pass to aircraft. It also contains a number of selectable pages with essential aerodrome information, radio frequencies, emergency checklists, rarely seen procedures and telephone numbers.

Lighting panel

This is a pictorial representation of the taxiways and runways on a 46in touch-screen. It is used to control the aerodrome ground lighting when it is switched on so that aircraft follow green routes embedded into the taxiways. At points where controllers want the aircraft to give way, there are red stop-bars that can be selected on the touch-screen. Heathrow is the only airport in the world to have a fully switchable aerodrome ground lighting system.

Runway incursion monitoring and collision avoidance system (RIMCAS)

The system is designed to monitor aircraft ten miles out from the threshold, and includes glide slope and localiser monitoring. As the aircraft nears the threshold, runway checks are performed and an alert sounds should another aircraft or a vehicle enter illegally. Having landed, distance, time and acceleration checks are performed against departures and any other objects on the runway strip.

Virtual control facility

At an undisclosed location in close proximity to the airport lies a virtual control facility that acts as a backup should there be a major emergency and the main tower becomes inoperable. It would take controllers less than two hours to be up and running to deal with flight traffic heading to and from Heathrow.

Heathrow simulator

To further improve the quality of training provided to controllers destined to work at Heathrow, NATS have installed a simulator to recreate the many changes planned for the airport. This is used for familiarisation, evaluation and training. A 360° 3D panoramic display shows a realistic view from the ATC tower. The cylindrical screen is 10m in diameter and combines ten powerful projectors and a continuous field of view. The simulator replicates daytime, night-time, dusk and dawn conditions with seasonal and inclement weather conditions, including fog, mist, rain and wind.

HEATHROW PEOPLE: JON PROUDLOVE

General Manager of Air Traffic Services

As I head into work my first thought always tends to be that at any given point in time there are around 200 different aircraft all pointing towards Heathrow.

On a typical day my first visit is to the office to check overnight emails, paying particular attention to any safety reports or observations. Then it's next door into the Operational Efficiency Cell for a briefing on the operation and the weather from the team. Once any queries or concerns have been addressed, I head to the lift for the ride up to the control tower. I always take the external lift as you can never tire of the view, though most mornings I remind myself that I must walk tomorrow – but do you know how many steps there are? Next there's a catch-up with the morning Watch Manager and a 'good morning' to the team in what must be the best office in the south-east of England.

With more than 1,300 aircraft movements a day and on average 200,000 passengers, it is essential that the day gets off to a good start, so the first few hours can often set the scene for the rest of the day. The airport is capacity constrained so there is no wiggle room, and this combined with external influences such as the weather can make our job quite an interesting challenge on some days.

We deal with a wide variety of issues, one example of which was a fire on an Ethiopian Airlines 777 parked on a remote stand. Fortuitously it was parked adjacent to the Fire Station, but from the tower the team could see smoke coming from the upper fuselage near the tail fin. Fortunately there were no passengers aboard the aircraft, which arrived earlier the same day and was not due to depart for another four hours. With almost all of the airport's firefighting resources dedicated to this incident we had no choice but to close both runways for 90 minutes. This resulted in a number of flights having to be diverted, but it wasn't long until things were back on track and the backlog cleared. It's incidents like this where the ATC really come into their own.

The majority of activity is based on precision-service delivery, with a clear focus on safety – keeping the operation safe is our number one priority.

CHAPTER 5

Terminals

OPPOSITE Heathrow's Terminal 5 complex – T5A (top), T5B and T5C (centre) viewed from the control tower at dusk. *(HAL)*

ABOVE Passengers collect baggage in the T5A main terminal building. The terminal features one of the most sophisticated baggage systems in the world. *(HAL)*

BELOW Terminal 5's interchange plaza connecting the multi-storey car park (left) with the main terminal building (right). *(HAL)*

Introduction

Heathrow's five terminals enable passengers to transfer between various forms of ground transportation and the facilities that allow them to board and disembark from aircraft – the essential link between landside and airside. In many respects the terminals are very much the 'public face' of the airport (and the airlines), as this is the environment passengers most commonly engage with on a regular basis. If passengers have a positive experience in the terminal it's likely they will become a repeat customer for the airport.

Within the terminal, passengers check in, transfer their luggage, transfer between flights, eat, shop, use the amenities, relax and pass through security. Heathrow currently spends more than £1 billion each year on its terminals,

making improvements to allow for quicker arrivals, smoother departures and reduced connection times.

Small cities in their own right, Heathrow's terminals must provide an environment which strikes a balance between the flows of people, the information with which they are provided to find their way through the airport, the needs of the airlines and the space available to supply services and amenities, all with the ultimate goal of getting passengers and their bags on planes in the safest and most efficient way possible.

Heathrow's investment in creating the world's largest integrated baggage system is at the heart of a £4.8 billion transformation, and this chapter sets out to explore how Heathrow's terminals operate and the various elements that make up a vital part of the airport's core infrastructure.

Terminal operations

With some 200,000 departing and arriving passengers passing through Heathrow's terminals every day, it's easy to understand why they are sometimes described as small cities. A terminal management team, led by an Operations Director and a Terminal Duty Manager (TDM) face a diverse and multi-layered challenge, encompassing all areas of the terminal, from passenger movements and security to offices and retail space.

Home for the TDM and his team is the Terminal Control Centre (TCC), which uses an extensive array of CCTV to achieve a full oversight of terminal operations. It is from here that the TDM can monitor departing and arriving flights, address passenger queries, make flight announcements and deal with any fire alarms or security breaches across the terminal's restricted zones. 'We use the control

centre as our eyes and ears on the ground for anything that is critical to the minute-by-minute operation of the terminal,' says T5's TDM Mark Coleman. 'The technology we have available enables us to make an early assessment of an issue and an informed decision about how it can be dealt with and ensure minimum disruption to passengers.'

A terminal the size of T5 has two TDMs, one of whom is based in the TCC and the other who is mobile, dealing with issues face-to-face. Given that T5 is a dedicated terminal for BA, the airline has its own Operations Centre adjacent to the TCC.

Successful terminal operations require the TDMs to specifically focus on managing the flow of passengers through security zones, liaising with airlines to support their efficient operation, providing special needs assistance to passengers and providing efficient, up-to-date information to passengers. The TDM also supports the terminal's retail and associated property businesses. The task of managing the terminals is all the more challenging given that the airport's infrastructure is operating at maximum capacity.

'As a regulated operation, Heathrow's performance in terms of moving passengers through the terminals is measured in 15-minute time slots to ensure we are meeting our target,' says T5 Operations Director, Susan Goldsmith.

'The airport is penalised if the targets are not met, so the TDM executes a series of what we call "operational interventions" to try and ensure minimum queue times for passengers.'

T5 deals with some 45,000 outbound passengers every day and strives to process around 35 passengers per security lane every 15 minutes, but a number of factors can increase or decrease this rate. A strong tail wind tends to get a flight in earlier, which means resources have to be on hand sooner to deal with passengers. As another example, in summer passengers are not carrying large coats and are generally wearing lighter clothes, making security checks easier. Conversely, large groups and a high proportion of families with small children departing on holiday can slow things down. The rate at which people pass through the terminal is closely monitored, and getting the right number of resources in the right place from a security perspective is one of the key balancing acts, particularly in T5, which has a north and a south concourse to deal with.

The time it takes passengers to pass through a terminal has a direct bearing on Heathrow's status as a hub airport, where minimum connection times are essential. 'We are constantly dealing with large numbers of people, and safety is our number one priority and of paramount importance in everything

BELOW An aerial view of the extensive Terminal 5 complex. The 'toast rack' layout is increasingly popular in large modern airports as it maximises use of an airport's land by placing the terminal building and its satellites perpendicular to the runways. (HAL)

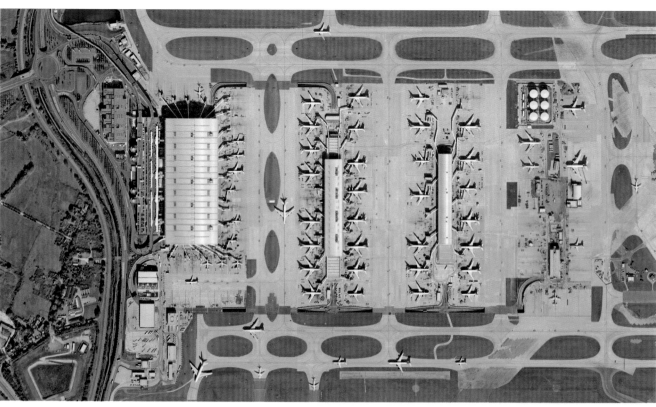

we do,' says Goldsmith, who encourages her team to adopt a balanced approach she calls 'courteous compliance'.

Terminal staff exercise continuously for emergency situations, such as establishing a friends and relatives reception centre in the event of a major crisis. This training will be enhanced and better supported in the future as the airport moves to aggregate the various TCCs as well as the STAR Centre into one, airport-wide master control centre. 'Better systems and analysis will allow for early warning of potential problems,' says Goldsmith, 'and this will make it far easier to get an overall assessment of an issue at the airport. From this we will better understand the knock-on effects of what is happening in one part of the airport and how it might impact elsewhere.'

With the last flight normally departing Heathrow around 23:30 at night, one might think the airport's terminals shut down, but in fact the window through until 04:30 is one of the busiest times in the 24-hour cycle. It's then that security equipment is serviced, engineering issues are resolved, cleaning takes place, retail outlets are restocked, and new signage and infrastructure are erected. Come 04:30 in the morning and the terminal's security lanes need to be manned, the shops need to be open and

LEFT **There's no place for traditional scaffolding as Terminal 5's interior roof structure gets a fresh coat of paint.** *(HAL)*

amenities fully functional to deal with the first passengers of the day.

With the advent of bigger aircraft such as the A380, most of Heathrow's terminals have had to be adapted to cope with this new challenge. This has meant considerable testing of air bridges, stand equipment, seating plans and gate layouts to accommodate increased passenger volumes – and this has all had to happen with no or minimal disruption to daily operations.

NEW TICKETING TECHNOLOGY

Heathrow is using pioneering technology across its terminals that will help improve passenger experience and departing flight punctuality. 'Positive boarding' will help reduce airlines' last-minute searches for passengers or their bags, as well as provide travellers with more accurate information to help them smoothly through their journey.

Computer software has been designed to be compatible with all airlines' computer systems. It enables airlines to see what stage of the departure process a traveller is at and gives passengers bespoke information to help them make their flight on time.

When the passenger presents their boarding pass, details from the bar code are compared against the central flight information, and tailored information for the individual flashes up on screen. For example, if a passenger is in the wrong terminal the message will tell them where to go, or if they have limited time, they are instructed to go through security and straight to the gate. If a passenger tries to go through with too little time before their flight is due to depart, they are asked to return to check-in and seek assistance from the

airline. This also allows the airline to begin unloading their baggage, as they have the 'real time' information of where passengers are in their departing journey.

BELOW **Self service check in facilities in Terminal 1.** *(HAL)*

Teams constantly carry out research to assess public perception and performance that can then be benchmarked against other European hub airports. 'Social media plays a massive part of our assessment process and allows us to see developing trends,' says Goldsmith. 'By aggregating our various social media feeds we can better understand an emerging issue and respond accordingly and learn how best to react if the same situation emerges.'

History of Heathrow's terminals

Terminal 1 (T1)

T1 was opened in 1968 and at the time was the biggest short-haul terminal of its kind in Western Europe. The terminal underwent substantial redesign and redevelopment in 2005, doubling the departure lounge in size and creating additional seating and retail space.

Since the buyout of British Midland International (BMI), British Airways service some short- and medium-haul destinations from this terminal, marking the airline's first return to this terminal since it moved out to T5 in 2008.

Despite upgrades of some £60 million in recent years – including self-service check-in technology and improvements to the baggage reclaim area – Heathrow plans to completely replace the building in 2019.

T1 facts
■ Caters for 13.8 million passengers on 123,000 flights annually.
■ Covers an area of 74,601 square metres.
■ Over 60% of flights go to UK and EU countries.
■ Connecting passengers: 26%.

Terminal 2 (T2)

Opened in 1955, the original Terminal 2 (T2) building was the airport's oldest terminal. Initially known as the Europa Building, it has seen some 316 million passengers pass through its doors. The terminal was designed to cater for 1.2 million passengers per year, but in its final years of operation it often accommodated around 8 million. The building was closed in November 2009 and subsequently demolished in 2010. The resulting space, combined with the area previously occupied by the Queen's Building, has made way for a new £2.5 billion state-of-the-art terminal which will eventually serve 20 million passengers a year. The terminal will open its doors on 4 June 2014.

Demolition of the original control tower that formed part of the central terminal area (CTA) of the airport began in January 2013, to make way for connecting roads. The new T2 will become home to 23 Star Alliance airlines, as well as Aer Lingus, Virgin Atlantic Little Red and Germanwings carriers.

Heathrow development director John Holland-Kaye says: 'The new T2 has been designed around the needs of our passengers, to allow them to get to and from their flights as quickly as possible.'

In June 2013 Heathrow announced that the new T2 would be officially named in the Queen's honour. Its full title will be Terminal 2: The Queen's Terminal.

T2 – BUILDING A NEW TERMINAL

Building and opening a new terminal is a huge challenge for any airport. The scale and complexity of the construction project and the number of contractors involved requires collaboration and co-ordination at every phase of construction, right up to the day of opening and beyond.

Like T5, much of the new Terminal 2 building has been constructed off-site, helping to overcome many of the logistical constraints of building in the world's busiest international airport. The new terminal's design continues the 'toast rack' principle employed in the construction of T5, a layout that maximises use of the airport's land by placing the terminal building and its satellites perpendicular to the runways.

The next challenge comes with moving the various airlines into the terminal – as space in general is tight at Heathrow, moving one airline has an effect on the operations of another, so each airline's move has to be planned in a specific sequence, factoring in the wider needs of each airline's ground handling staff who look after baggage and customer service.

To ensure the opening of the new terminal is successful, a number of key activities are undertaken, including:

- **Research** – looking at every terminal opening that's taken place in the past ten years and examining the problems they had, to see what can be learnt.
- **Co-operation** – the airport has worked continuously with the incoming terminal occupants, using working groups to cover every aspect, from check-in and airline moves to baggage handling and staff logistics.

- **Construction** – a rigorous planning and build schedule has been kept on track, allowing a six-month window before opening for testing and staff training.
- **Operations Director** – Heathrow appointed the terminal's Operations Director 18 months in advance, thereby helping to keep construction decisions focused on safe operations and improved passenger experience.
- **Trials and test events** – the six months between construction handover and terminal opening allow the terminal's management team to plan, test, improve and test again before opening for business. More than 180 trials have tested all aspects of the terminal. Larger trials have involved more than 3,000 people at a time.
- **Baggage handling** – the baggage system has to be tested on numerous occasions at full capacity to replicate real-life operations.
- **Staff training** – extensive operational and technical training to ensure staff know how the new terminal works and understand new systems and processes.

T2 will open in phases, with the 26 airlines moving in over a period of six months. This means the terminal will start out with a limited number of passengers as it builds up to full operations, allowing the team to iron out any teething troubles. This will ease the transition and allow the airport to avoid some of the challenges faced when T5 opened in 2008.

BELOW **Gate seating areas and the raised flight arrivals walkway prior to the opening of Satellite Terminal 2B.** *(HAL)*

ABOVE A Pan Am
Boeing 747-100 'Clipper
Defender' on stand at
Terminal 3 in the 1970s.
(HAL)

dedicated belts to cope with the large luggage
volumes associated with A380 flights.

T3 facts

- Caters for 19.4 million passengers on 104,000
  flights.
- Covers an area of 98,962m^2.
- More than 45% of flights go to non-EU countries.
- More than 33% of passengers are visiting
  friends and relatives.

Terminal 4 (T4)

T4 was opened in April 1986 and for some 22
years served as a primary base for British Airways
(BA) before the airline relocated its operations to
the new Terminal 5 in 2008.

The terminal's distance from the central
terminal area made passenger and baggage
transfers complicated, although this problem
was somewhat alleviated in the late 1990s by
the construction of an automated transfer tunnel
between the CTA and T4. Following the departure
of BA the terminal has undergone a significant
transformation with the investment of some £185
million since 2007, and now plays host to 45
airlines and is the current base for the 'SkyTeam'
airline alliance, including Etihad Airways, Malaysia
Airlines, Gulf Air and Qatar Airways. Recent
upgrades include new stands to accommodate
the Airbus A380 and a new baggage system.

T4 facts

- Caters for 9.9 million passengers on 62,000
  flights.
- Covers an area of 105,481m^2.
- Over 73% of flights are to non-EU countries.
- Focused on direct passengers (only 12.7%
  transfer to another flight).

Terminal 3 (T3)

Opened as the 'Oceanic Terminal' in 1961,
Terminal 3 was originally built to handle flight
departures for long-haul routes. It was expanded
nine years later with the addition of an arrivals
building. In 2006 the new £105 million Pier 6
was completed to accommodate the Airbus
A380s operated by Singapore Airlines, Malaysian
Airlines, Emirates and Qantas. A year later saw
the addition of a new drop-off area and a large
pedestrianised plaza.

Further upgrades worth some £1 billion are
planned for the rest of the terminal over the next
ten years, which will include the renovation of
aircraft piers and the arrivals forecourt. A new
baggage system connecting to T5 is currently
under construction and additional improvements
are planned for the baggage claim hall, with

BELOW Terminal 3's
refurbished forecourt.
The terminal plays
host to major airlines
including Emirates,
Air India and Virgin
Atlantic. (HAL)

Terminal 5 (T5)

Opened in 2008, T5 is the largest free-standing
structure in the United Kingdom. At nearly
400m long, 175m wide and 40m high, the main
terminal's roof area is the size of five football
pitches. The terminal was used exclusively by
British Airways until March 2012 but has since
become a hub for BA's owners, the International
Airlines Group (IAG), which includes the Spanish
national carrier Iberia. T5 was the first new terminal
at Heathrow for a quarter of a century and serves
more than 30% of the airport's passengers.

The building cost £4 billion and took 19 years
from conception to completion, including the

LEFT A 2006 aerial view of the Terminal 5 construction site – the project included 60 new aircraft stands, two satellites terminals, a 4,000-space multi-storey car park, a new spur road from the M25 motorway and more than 13km of bored tunnels. (HAL)

longest public inquiry in British history. Construction began in September 2002 and just three years later the final section of the main terminal roof was lifted into position. Interestingly, the roof could not be raised with conventional cranes because it would have penetrated vertically into the airport's radar field and it had to be assembled on the ground using smaller cranes and then lifted into place by eight custom-built towers.

Unlike most airport terminals, the main terminal building does not have direct road access and is instead fronted by a multi-level building containing a bus station, taxi rank, multi-storey car park and passenger drop-off zone. More than 15,000 volunteers participated in 68 trials to test the operational readiness of Terminal 5 prior to its opening by Queen Elizabeth II on 14 March 2008. Despite the trials, the new terminal suffered a number of initial operational challenges with more than 500 flight cancellations and some 42,000 bags that failed to travel with their owners. These issues were soon overcome and T5 today ranks amongst the best airport terminals in the world.

The main terminal building is referred to as Terminal 5A. There are two satellite buildings, Terminal 5B and Terminal 5C. An underground automated people mover (APM) system transports passengers between Terminal 5A, Terminal 5B and Terminal 5C. The system can accommodate up to 6,800 passengers per hour.

T5 facts
■ Caters for 26.3 million passengers on 185,000 flights.
■ The main terminal covers an area of 300,000m^2. The two satellite terminals cover a further 120,000m^2.

■ Over 50% of flights go to non-EU countries.
■ Almost 50% of passengers connect to another flight.

Royal Suite and VIPs
Tucked away on the south side of the airport is Heathrow's Royal Suite, used for 'red carpet' movements of members of the royal family, visiting heads of state and some of the world's best-known celebrities. Built in 1991, the terminal has its own stand and is specifically designed to cater for VIPs; it is only opened on special occasions. In addition to the Royal Suite, the airport also has a number of VIP suites, used for royalty, diplomats and politicians who are travelling on regular passenger services. Known as the 'Windsor Suite', 'Hounslow Suite', 'Spelthorne Suite' and 'Hillingdon Suite', they are located in or near different terminals.

BELOW Heathrow's Royal Suite has been welcoming VIPs and heads of state for decades. (Author)

Baggage

The art of baggage handling is a challenging one and getting it right is fundamental to keeping passengers happy. The airport has to deal with millions of passengers each year, manage peak and off-peak passenger levels, connect passengers through Heathrow's hub to other flights within minutes, adhere to rigorous health and safety requirements, keep all items secure, transport baggage from the terminals to the aircraft quickly and ultimately keep track of millions of units of luggage.

The measure of a successful baggage-handling system is simple: can the bags move from point to point as fast as the passengers can?

When passengers check in at the terminal, luggage tags are printed and attached to each piece of luggage. The tag has all of the passenger's flight information on it, including the destination, stop-over cities and a bar code that contains a ten-digit number. This bar code and number are unique to the item of luggage, and is used to track the bag at any point along its journey.

THIS PAGE Terminal 5's sophisticated baggage handling and distribution centre. There are 30 miles of conveyors, 2.8 miles of tunnels, 44 baggage reclaim belts and around 53 million bags processed every year. *(HAL)*

After check-in, the bag passes a series of automated bar-code scanners arranged at 360° around the conveyor (including underneath). This device is able to scan the bar codes on about 90% of the bags that pass by. The rest of the bags are routed to another conveyor to be manually scanned. Once the baggage-handling system has read the ten-digit bar code number, it knows where the bag is at all times.

The high-speed conveyor system then routes the bag through different security scanning devices, and, once cleared, high-speed 'diverters' read the tag and send it down a particular route so that it ends up in the relevant airline's baggage make-up area. Bags are then either loaded on to luggage carts and transported to the aircraft for bulk loading – meaning the bags are brought up one by one on a conveyor and placed into shelves in the cargo hold – or they are loaded into containers on the ground and then placed into the hold of the aircraft. When loading the plane, bags belonging to transfer passengers are loaded into separate areas from bags that will be heading to baggage claim. A monitor at the sorting station tells the handlers which bags are going where.

Since Heathrow is a major hub, a large percentage of passengers coming through it are transferring to other flights, sometimes from a different terminal to the one they arrived in. Again, the goal of the system is to have the bags keep up with the passengers. Generally speaking passengers disembark faster than the bags can be unloaded, so for the bags to keep up they need to be able to move quickly between gates and terminals.

Baggage is removed by the airline's ground-handling agents as part of the turnround procedure and loaded on to carts that are pulled by tugs to the baggage claim area. Since the bags are already sorted when they come off the plane, it is easy to keep the transferring bags separate from the terminating bags. When the bags get to the baggage-claim area they are loaded on to a short conveyor that deposits them on to the carousel. Oversize items like bicycles, golf bags and large souvenirs are routed to a special out-size carousel.

In T5 alone, the baggage system has more than 30 miles of conveyor belts driven by 10,195 electrical motors. The system is capable of handling 70,000 bags per day, although a realistic daily average is more in the order of 50,000.

DEVELOPING AN INTEGRATED BAGGAGE SYSTEM FOR HEATHROW

Given Heathrow's vision to be Europe's hub of choice, in which passengers and airlines think of Heathrow as the airport where bags move with speed, efficiency and certainty, development has commenced on a Heathrow-wide integrated baggage-handling system capable of handling 110 million bags a year. Once operational, this will enable the airport to process the departure and transfer baggage for passengers more efficiently, and thereby minimise transfer times for passengers on connecting flights. It will also serve to reduce bag misconnection rates, allow shorter build times for flights and will provide enhanced management information.

The first stage of this mammoth project is to link T5 and T3 via a tunnel and the Western Interface Building (WIB). The connection creates a single baggage system that extends T5's world-class baggage handling into the Central Terminal Area.

Inter-terminal connections between T5 and T3 will be automated by a baggage tunnel, eliminating roughly 120,000 vehicle movements a year from the Heathrow road network. The tunnel will carry up to 3,000 bags an hour each way. With more predictable and shorter connection times, airlines using Heathrow will be in a better position to offer more competitive connection schedules, making the airport more attractive for connecting passengers.

'Manual-handling aids will ease the physical burden of baggage handling, especially for staff who load baggage containers,' says Stephen Livingstone, Western Baggage Product Programme Director at Heathrow. 'We'll ease the pressure on everyone by building flights early, using automated and semi-automated equipment. Because the bag store can sort bags by size and weight and in convenient batches, handlers will know what's coming. The flow of bags will be more predictable and the quality of work will be better.'

Ultimately everyone benefits from improved performance: passengers fly with more confidence knowing that their bags are travelling with them, airlines have better knowledge of where bags are in the system, and handling agents enjoy improved efficiencies and more predictable workflows.

BELOW **Heathrow's existing baggage system is in the midst of a major overhaul.** *(Author)*

Signage

Wayfinding and signposting across the airport plays a vital role when it comes to leading passengers on their journey through a terminal, particularly those who are not frequent travellers. 'Some travellers, for example, arrive at T5 but don't necessarily understand the fact that there is a main terminal building and two satellite terminals which are connected by train,' says T5 Operations Director Susan Goldsmith. 'If this is not communicated correctly and passengers are not provided with a sense of the time it will take to get to their gate, we create unnecessary delays and reduce efficiency within the terminal.'

Heathrow uses a variety of different wayfinding aids – some design devices are used directly to show the way, like signs or maps, while others can be used indirectly, such as lighting, flooring,

prominent landmarks or architectural features. 'Whenever possible,' says Goldsmith, 'Heathrow's terminal facilities are designed around logical traffic patterns that enable passengers to move easily from one point to another, on an intuitive basis, and without confusion.'

Wayfinding is a series of interlinked decisions, each of which affects the outcome of the next one. 'Where possible, we try to limit the number of decision points that a passenger encounters, to make their journey as seamless as possible,' adds Goldsmith.

There is a continual evolution of the airport's signage, which is carefully considered in relation to the physical design of the terminal. Good architectural design of an entrance visually emphasises it much more effectively than a sign stuck above a door. Clearly defined and well-lit pathways will lead people much more intuitively than a directional sign, however well designed. Wayfinding has been boosted by the availability of free Wi-Fi in the terminal, allowing passengers to make better use of mapping applications and the airport's own App.

Signage also helps to reassure passengers, as they move through terminal buildings and reach key decision points, that they are in the right place and heading in the right direction.

Signs within the Heathrow system feature a range of standard white-on-black pictograms. These have been carefully developed through passenger research and provide a universally understood visual language. These pictograms are combined with black lettering using a consistent and easily understood font on a yellow background, which has been proven to be the most legible combination for both lit and unlit signs. White text out of purple is used to distinguish and identify airside passenger connections, routes and facilities. English, the most internationally recognised language, is used for the sign wording and the vocabulary has been standardised to ensure that the words used have maximum clarity and meaning internationally.

HEATHROW AND THE LONDON 2012 OLYMPICS

Heathrow was nominated as the Host Airport for the London 2012 Olympic Games and Paralympic Games and went to great lengths to welcome the world. Preparing the airport for the Games required a huge team effort, and in the lead-up to the Games the airport established 'Team Heathrow', a core group of organisations including airlines, ground handlers, police, immigration staff and LOCOG to ensure the airport was capable of delivering an outstanding welcome to thousands of passengers – athletes, officials, fans and the global media – and their baggage.

Some 1,000 volunteers relentlessly tested the terminals before they were dressed with Games signage and the airport was declared ready to welcome the world.

Leading up to the Games, a dedicated Olympic Terminal was erected at Heathrow near T4. Athletes were offered the opportunity to check in at the Olympic Village in London and from there they could travel by bus to the Games Terminal. To give a sense of scale, the terminal roof weighed 64 tonnes and took four cranes two days to raise it. To cope with the demands placed on the airport, a 'Games Control' team was established, with representatives from every operational area acting as a single point of contact for Heathrow's Games operation.

The team was operational for 83 days, each of which was assigned a status – green, amber or red – depending on the number of passengers expected through the airport. There were 18 red, 23 amber and 42 green days.

Special media centres were set up in each terminal to cope with extensive global interest in the Games. While journalists were writing their copy, athletes were arriving from every quarter with a wide array of equipment including bicycles and canoes. When the Olympic torch finally departed London, figures showed that more than 500,000 people had flown in for the Games and Heathrow had coped more than admirably with the challenge.

BELOW Construction of the London 2012 temporary Games Terminal. *(HAL)*

ABOVE High-end eateries cater for all tastes. *(HAL)*

Retail

It's no surprise that with over 52,000m² of retail space and more than 340 retail and catering outlets, Heathrow offers an unparalleled shopping experience which ultimately translates into a critical revenue stream for the airport. With everything from luxury fashion to consumer electronics and beauty services to shoe shines, shopping at Heathrow has almost become a much-anticipated part of the journey for many passengers. The recent uplift in luxury sales at the airport has been driven in part by the growth in disposable income in emerging markets such as Brazil, Russia, China and India.

Airports represent the beginning and the end of a journey, special emotional states for passengers, and the retail environment can be the solution to travel stress. Heathrow's retail offerings are designed to add to the passenger experience, but this has to be balanced against the space needed in the terminal for basic operations, facilities and seating.

Retail facts and figures
- Heathrow sells over 1,050 bottles of champagne every day.
- Heathrow sells 26,000 cups of tea and 35,000 cups of coffee every day.
- Heathrow sells 700 muffins and 1,800 sandwiches per week.
- Heathrow sells 990,000kg of chips every year.
- More than 100,000 bottles of fragrance are sold per week.
- One bottle of Chanel Nº5 is sold at World Duty Free every nine minutes.
- The amount of pastries Heathrow sells each year would line the runway – in both directions – 350 times.
- If you stacked all the loaves of bread that the average Heathrow sandwich shop uses each year, it would be 35 times higher than the Heathrow air traffic control tower.
- Breakfast is the most popular dish of the day at Heathrow, with almost 5 million eggs, 6.4 million croissants and a whopping 4.5 million rashers of bacon served every year.
- The airport dishes up more than 1.5 tonnes of caviar, 25,000kg of salmon and over 240,000 oysters annually. This delicious seafood requires a garnish of 400,000 lemons every year.
- Heathrow pours over 800,000 glasses of champagne every year.
- Over 250,000 pieces of sushi are served at Heathrow each year, and 630,000 portions of steak.

Advertising and media

Another one of the airport's most important revenue streams comes from advertising and media, much of which appears in the terminals. More than 1,200 different advertising sites across the airport – many of them digital – are managed by one of the world's leading advertising companies, JCDecaux.

The demographic profile of Heathrow's passengers makes the airport a tremendously attractive location for advertisers. Industry-leading business travellers are open to information and distraction, and affluent leisure travellers are looking to spend both their money and time. 'With 90% of our passengers coming from an AB demographic, Heathrow's advertising spaces are highly sought after by brands looking for exposure to an international marketplace,' says Nick Webb, Heathrow's Head of Advertising, Sponsorship & Expo.

Consumers expect to see cutting-edge technology as part of the airport media they consume, providing advertisers with a welcome channel through which to create deeper engagement and build a relationship with this valuable and elusive audience.

The key categories of advertisers include the transport and travel sector, followed by luxury goods, electronics, computer software and services. There are a small number of unique and 'iconic' sites created specifically to the requirements of advertising clients within the central terminal area, including the 'Emirates roundabout' and the 'Turkish Airlines globe'.

Surface access

It's all well and good being able to move passengers through the airport, but provision has to be made to get them to – and from – Heathrow efficiently. To this end, there are a number of key components of Heathrow's surface access infrastructure, including:

■ Three rail stations: Heathrow Central (for T1, T2 and T3), T4 and T5. The airport also has three

ABOVE Buses and taxis form an integral part of the Heathrow transport network. *(HAL)*

RIGHT Heathrow's Personal Rapid Transport vehicles provide passengers with a flexible and environmentally friendly transfer solution. *(HAL)*

RIGHT The airport offers charging bays for electric vehicles. *(HAL)*

stations on the London Underground, one for T1, T2 and T3, one for T4 and one for T5.

■ Car rental facilities, drop-off, valet facilities, taxi and coach feeder parks.
■ A central bus station located between T1, T2 and T3.
■ Car parks that provide more than 22,900 spaces.
■ Of those passengers that depart directly from Heathrow, research shows that:
■ 30% arrive in a private car, and of this total around 56% are dropped off on the forecourt.
■ 18% drop off in short-stay car parks.
■ 19% park in long-stay car parks.
■ 7% park off-airport.
■ 12% use a scheduled bus or coach service.
■ 25% use a taxi.
■ 27% use a rail service.
■ The rest come by hotel shuttle, charter bus or hire car.

Heathrow Express

The Heathrow Express non-stop rail service was launched in 1998 as a premium high-speed rail link, taking passengers from Paddington to Heathrow in just 15 minutes, the fastest route to central London.

PRT

Heathrow has introduced a personal rapid transport (PRT) system to provide a sustainable alternative to traditional bus, coach and car use at the airport. The system links the T5 business car park to the main terminal building and is composed of a fleet of low-energy, battery-powered driverless vehicles capable of carrying 1,000 passengers and their luggage daily along a dedicated guideway. The system generates zero local emissions and is 70% more energy efficient than traditional airport buses. In just one year of use, the system alleviated the use of 50,000 bus journeys and the corresponding CO_2 emissions.

Parking

Approximately 8 million customers utilise official Heathrow parking each year – some 430,000 in long stay and 7.6 million in short stay. The airport's parking facilities provide passengers with secure and monitored parking facilities within the airport boundary. The airport also provides free charging facilities for passengers who own electric cars.

T5 Terminal Duty Manager

As a TDM, my main task revolves around looking after some 50,000 customers a day, managing up to 600 security officers, overseeing 30 front-line managers, or Service Team Leaders (STLs), making sure our terminal is safe and secure, putting passengers at the heart of everything we do and, finally, instigating and leading contingencies that may arise at any time.

Of critical importance is ensuring all equipment and building systems are working correctly after mandatory maintenance carried out the night before. T5 is a huge facility, so it's not uncommon to have several maintenance teams on site at night.

Making sure we have enough security staff to deal with the expected volume of passengers is also high up on my priority list. All the search areas have to be opened at the allocated times and I spend a fair amount of my time each day reviewing how our security staff are allocated and then making changes when needed for the benefit of our passengers.

Given its size, T5 has two TDMs – one mobile in the terminal and one static in the Operations Centre. It's imperative we're on top of all the key issues of the day and I always dial into the airport's daily conference call to update the Airport Duty Manager of any key issues affecting T5, but also to understand what is going on across the rest of the airport, as this could have an impact on the terminal and passengers, particularly those transferring from other flights.

Incidents in the terminal are few and far between but when they happen the entire team pulls together to address things quickly. As an example, we recently had a fire alarm activated in a food retail outlet. Four service team leaders were deployed to the incident to complete a risk assessment for signs of smoke or fire. The Airport Fire Service and London Fire Service attended to assess the situation and declared the area needed to be evacuated. Following a careful assessment, there was fortunately no fire risk and once the Fire Service was happy the area was safe, the teams stood the incident down and passengers were able to return to the area.

BELOW One of the TDM's first priorities is to ensure passenger needs are met. *(HAL)*

CHAPTER 6

Security

OPPOSITE A police vehicle passes in close proximity to an Air New Zealand aircraft parked on stand at Terminal 3. *(HAL)*

Introduction

For more than 40 years the aviation industry has had to counter and respond to a variety of security threats, with airports and airlines seen as high-profile targets for terrorists seeking to publicise their cause on the international stage. Airports are also seen as significant targets due to the economic impact of attacks against them, and the fact that a nation's representatives, such as national flag carriers, can be targeted.

The high concentration of people at Heathrow – around 70 million passengers passing through the airport each year – and the iconic nature of this global crossroads can appear to provide an alluring target for terrorism. For the security team at Heathrow, one of the key tasks is to counter terrorists' tried and tested methods as well as to anticipate new ones. Step back in time to when Heathrow started operating and people were allowed on to the apron tarmac to welcome friends and family on an inbound flight. There was no perimeter fence to speak of. In later years a fence was erected, but this was more a health and safety response, to stop people wandering on to the airport. Today, by contrast, there is a significant fence – reinforced to resist penetration – driven by security requirements, as part of a substantial security infrastructure to protect the airport, its staff, its passengers and its assets.

Today Heathrow needs to be resilient in the face of all hazards, ranging from wide-scale health concerns to climatic events. Security threats are no exception. Security has become a major factor of spending for the airport, and the need to adapt to and implement security policy introduced by the government in response to increasingly innovative threats. 'The security situation is very fluid and we are required to respond to all the prevailing risks,' says Heathrow's Security Director, Francis Morgan.

Ultimately, an effective security operation is fundamental to maintaining Heathrow's operational capability. While the airport needs to maximise overall safety and security, it must also improve the passenger experience and deliver high levels of operational efficiency.

With airport security a highly sensitive topic, regulations understandably prevent this chapter from being too descriptive. Approval had to be sought from a number of different government agencies and departments prior to publication.

Despite some of the limitations, however, this chapter seeks to explore the various techniques and technology used in protecting passengers, staff, aircraft and the huge array of facilities and infrastructure at Heathrow. It also looks at typical threats the airport must be prepared for and the demands of meeting government security requirements in a post-9/11 world.

Aviation security regulation

A series of international and government bodies are responsible for formulating aviation security policies, from a global level to the airport itself. These ultimately define airport security practices and invariably have a direct bearing on passengers passing through the airport.

In the first instance, the International Civil Aviation Organization (ICAO) sets global standards and recommended practices designed to prevent and suppress acts of unlawful interference against civil aviation. Its aim is to set up a common baseline to ensure the security of all flights.

The European Commission (EC) works to establish common rules and standards on aviation security across the European Union (EU). EU legislation applies to all civil airports located across Europe. The EU regulations also require member states to designate a single authority responsible for the co-ordination and monitoring of aviation security. In the United Kingdom this falls to the Department for Transport (DfT). The EU, together with a further dozen European countries, form a free market in aviation. Airlines within this space can fly passengers freely within the area, so agreed minimum security rules are important.

In addition to drawing up and implementing a national aviation security programme, the DfT takes the lead when it comes to introducing and enforcing more stringent security measures to combat nationally identified risks. As the national hub airport and a key part of the country's critical infrastructure, the government maintains a key interest in all aspects of Heathrow's security policy.

There are various other formal structures, including the National Aviation Security Committee (NASC), which provide the opportunity for senior aviation representatives from government, the police and the wider aviation industry to discuss security matters at a strategic level.

The recommendations from these forums are incorporated into the broader aviation security programme to be implemented at airports across the country and a bespoke Airport Security Programme for Heathrow.

From 2014 some of the DfT's functions, including regulation, inspection and enforcement of security matters, will be transferred to the country's Civil Aviation Authority (CAA) to allow for improved efficiencies.

Overview of security at Heathrow

'Security touches everything that Heathrow does,' says the airport's Aviation Security Manager. The passenger search is just one part of a complex and robust security programme that starts with security as a key consideration in the airport's overall design. 'In the construction process we go to great lengths to make use of the right materials to withstand an explosion and minimise its effects. Car parks are kept a minimum distance away from the terminal itself, the flow of people is designed not to create targets, and even the airport's road layout is designed to stop vehicles from building up speed.'

Next on the list of key considerations are Heathrow's critical and strategic assets that are vital to the safe and continuous operation of the airport. These include the aircraft, terminal buildings, power supply facilities, runways, lighting systems, fuel storage and air traffic control tower to name but a few.

Potential threats to the airport also come in various forms, many of which are unconventional. To provide a sense of the challenges involved, consider that airside roads must have restricted access and be used only by authorised vehicles performing specific functions. The airport needs contingency plans for vital services such as electric power, telecommunications, water and transportation. Security personnel at Heathrow need to be trained to deal with a variety of risks, respond to incidents, work with the latest technology and be educated to analyse complex situations. The airport needs fast response teams in place in critical areas to provide both visual and covert security protection. Personnel and supplies entering secure airside areas have to be screened. The airport's information systems and network architecture have to be secured against unauthorised use or access.

For security purposes, the whole airport is divided into two areas, airside and landside. The difference between the two is that everyone airside – which is classified as a restricted zone – is there

BELOW Onion ring philosophy: Heathrow and its assets sit at the core of the 'onion', while the layers of 'onion peels' are security systems and deployments put in place to protect the airport. The concentric rings act as protective layers, each one difficult to breach on its own, more so when combined with the others. *(Roy Scorer)*

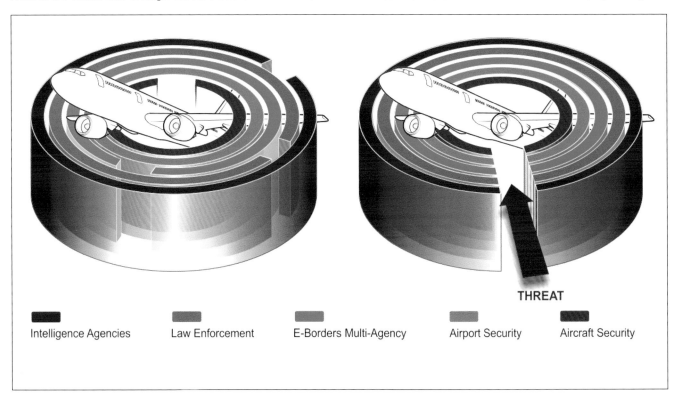

THREAT

Intelligence Agencies   Law Enforcement   E-Borders Multi-Agency   Airport Security   Aircraft Security

CHANGES IN AIRPORT SECURITY SINCE SEPTEMBER 2001

Air travel has changed more in the last ten years than in the half-century that went before it. When Heathrow started out life in 1929 as a small airfield few people would have guessed just what a prominent role security would come to play in the airport's future development.

Even before the 9/11 terrorist atrocities, it is evident that the level of security has had to evolve together with the airport – when, for example, the airfield was used in the mid-1940s by long-distance military aircraft bound for the Far East an increased level of protection was required.

The most telling changes came about in response to 9/11 and the successfully disrupted attacks in 2006 that prompted a massive overhaul of airport and airline security. Almost overnight the aviation industry was implementing locked cockpit doors and air marshals.

The liquid-bomb plot in 2006 had a significant impact on aviation security in the UK and across the world. To combat the potential new threat a range of restrictions came into force in August of that year, including an initial prohibition of all carry-on luggage (except essential items such as travel documents and medication), although this rule was later relaxed to allow hand baggage again and liquids in small quantities, including liquid medications and baby milk, if they were tasted first by passengers at the security checkpoint.

The 2006 plot forced governments, airlines and airports to identify different ways of delivering security and Heathrow was at the forefront of this change. Such sudden and dramatic increases in security measures come at significantly increased cost to the airport, require the employment of additional resources, invariably take up valuable space within and around the terminal buildings and add to the complexity of delivering a positive passenger experience. More than ten years on from the 9/11 attacks and many people question whether conducting uniform security searches of millions of passengers who are unlikely to pose a threat is the best approach, preferring airports to adopt a passenger screening regime that gives preferential treatment to low-risk passengers. Others argue that attempting to identify high-threat groups could underestimate the range of terrorist and criminal groups. While a change to the current approach will inevitably be the case (see 'The future of airport security' later in this chapter), it is fair to say that Heathrow's security system is robust, and current risks have been mitigated effectively by new measures.

because they are either travelling or carrying out important services for the airport's operation, and has been subject to a security search and has a reason to be there, either because they are travelling or because they are supporting the operation in some way. For passengers, this is defined as having shown their boarding card and been through the security search. Staff and their vehicles have their IDs checked and are searched just like passengers.

Landside areas, by contrast, are all the areas of the airport open to the public, including transport and terminals before the security search.

Heathrow's police team

The Metropolitan Police are a fundamental part of the overall team that co-operates to deliver security at the airport. The other key players are Border Force, airline security and intelligence services. These groups combine to make the airport welcome to the travelling public while at the same time deploying effective, reliable systems and procedures to deter and counter a range of threats.

Heathrow Airport is policed by a Metropolitan Police Service Operational Command Unit (OCU) known as SO18 Aviation Security, which operates from an on-site police station with more than 400 active officers on duty. The role of the Metropolitan Police Service is to keep Heathrow Airport safe for the travelling public, staff and visitors. This is done in partnership with other law enforcement agencies such as the Border Force, the Serious Organised Crime Agency (SOCA) and, soon, the National Crime Agency (NCA).

BELOW A police dog handler and his dog carry out a security search in Terminal 1. *(HAL)*

RIGHT An armed police patrol in Terminal 5. *(HAL)*

Aviation Security is an armed OCU, which means many officers carry firearms, although there are several other teams within the command. The airport also has a dedicated team of officers within the terminals to handle local community issues.

The OCU deals with the threat from terrorism, the protection of Heathrow as an iconic site and the response to organised crime and volume theft on a daily basis. The unit works closely with the other emergency services and Heathrow to provide a high level of response to aircraft emergencies.

In the event of a major crisis at the airport, a diverse team of support staff and specialists are on standby to resolve the situation. This includes operational command and control, air traffic control, explosive ordinance disposal, armed intervention teams, interpreters, hostage negotiators, the police, fire brigade and ambulance services.

Perimeter fence and control posts

At 4.7 square miles, Heathrow is a massive piece of real estate to keep secure, and its 9.5-mile perimeter fence – much of which runs alongside public roads – forms what effectively amounts to a 'first line of defence'.

The perimeter fence has the dual role of deterrence and detection – it helps to minimise the likelihood of unwanted intruders breaching protected landside and airside areas. Security patrols regularly scan the perimeter in case someone tries to cut through the fence. The fence itself is a real deterrent to entry – it is high and of a non-scalable metal construction. Along the perimeter care has been taken to secure all

IRA MORTAR ATTACK IN MARCH 1994

Heathrow's perimeter fence was breached in an unconventional way in 1994 when high-explosive mortar shells fired by IRA terrorists from a nearby hotel landed on the apron of the northern runway. In a second attack the next day four shells fired from a secluded spot in a patch of woods next to the A30 struck an area where planes park at Terminal 4. Neither attack caused any injuries, but both disrupted flight traffic.

LEFT An airfield security observation post. (HAL)

LEFT A member of the security staff surveys the airfield with binoculars. (HAL)

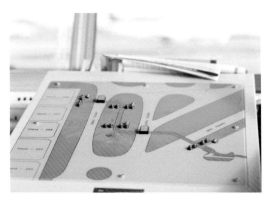

LEFT Taxiway warning light control module. (HAL)

ABOVE AND BELOW Heathrow's control posts provide secure access for vehicles and personnel. *(HAL)*

conduits, sewers and pipes to ensure that entry to the airside is not possible in this way.

Around the fence line there are 18 official control posts, the main purpose of which is to provide controlled access into the airport for everyone but passengers. These posts are manned 24 hours a day by security personnel, and anyone entering the airport premises through one of these gates is subject to the same search as a passenger passing through one of the terminals.

Personnel

Interestingly, almost all airport security is done in-house by full-time Heathrow employees, who make up more than half of the airport's 6,500-strong workforce. Some airline security services are outsourced to special third party providers and these personnel are subject to strict vetting procedures.

Heathrow operates a sophisticated electronic access control system for all personnel on site, with access to all areas of the airport graded according to individual job requirements. The control system is designed to efficiently manage both the issuing and cancellation of access cards. The system also serves to prevent and control the movement of persons and vehicles to and from security-sensitive areas of the airport property.

The system has to ensure that each employee has access to the areas they need and no more and is designed to log, authenticate and track personnel movements in and out of key areas, producing a real-time audit trail of contractors, maintenance crews and permanent staff.

In addition to Heathrow's own employees, the security system has to cater for up to 70,000 people on site every day including airline employees and crew, retail staff and handling agents. Furthermore, refurbishment work at the airport invariably means large numbers of external contractors on site, and the access system serves to keep control over their movements too.

'Everyone who works at the airport, from check-in to cleaning, to PR and catering, have a role to play in the security – from challenging suspicious behaviour, knowing the operational procedures for evacuation of large groups of people and IT security for sensitive material,' says the airport's Security Risk Manager.

Passengers

The airport's first priority is the safety and security of people, and there are a lot of them – some 72 million passengers travel through Heathrow every year.

To cope with this, Heathrow has a comprehensive layered security system that protects travellers and is also used to assist in managing passenger throughput within the airport. This is important for a number of reasons: for the passenger, reduced queuing times contributes to a more pleasant airport experience, while for the airport, it helps to optimise staffing levels and increase the time passengers spend in the increasingly important retail areas. The airport has the challenging task of balancing the passenger experience – making it as intuitive and engaging

LEFT Hand luggage security search trays. *(HAL)*

as possible – with mandated security requirements that set uncompromising standards.

To this end, Heathrow has introduced new initiatives designed to improve the security experience. 'Family friendly' lanes make life easier by assisting mums and dads with energetic toddlers to prepare for security. John Walker, Head of Operations at Terminal 1, says: 'Clearing security for a family can be tough, so we try to make it as simple a process by giving the kids subtle distractions by way of engaging graphics and displays to enhance what is a very sterile environment. This makes for a better all-round experience and gives families more time to prepare.'

Another innovative step forward is the use of automatic ticket presentation gates that improve the passenger's journey through the airport. 'People feel more enabled, and happy passengers equal happy customers,' says Walker. There are also improved spaces for passengers to sit and reconcile their belongings after the security check is complete.

With passengers arriving from and departing to more than 90 countries, consistent and effective security signage plays a vital role. 'One of the key challenges in security from a passenger perspective comes where there is a language barrier,' says Walker. 'The airport needs to overcome the language barrier by making practical use of real life examples to demonstrate what is and is not permitted.'

The airport has also tried to form a subconscious link through the design of the signage, with black/red signs delivering a security message as opposed to black/yellow signs used for wayfinding.

Improved signage helps to process passengers at a rate that allows Heathrow to meet set

LEFT Security staff pay particular attention to liquids carried in hand luggage. *(HAL)*

BELOW A member of the security team reviews passenger boarding cards. *(HAL)*

customer service targets, failing which penalties can be imposed by the CAA. To achieve this, the security lanes are closely monitored, and much time is spent studying passenger peaks and

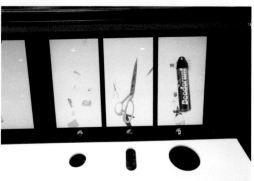

troughs to allow for the best possible allocation of
resources and a balance across the airport.

Technology has always been and will continue
to be integral to security, helping to improve
efficiencies and allow the airport to process
passengers more quickly. Some of the state-
of-the-art technology that sits at the heart of
Heathrow's security system includes:

■ More than 6,500 CCTV cameras monitoring the
entire airport.
■ Passengers pass through metal detectors
that create a brief magnetic field. If an item is

found, passengers are asked to remove any
metal objects from their person and will be
hand searched. Security staff may also use a
handheld detector based on similar technology
to isolate the cause.
■ Carry-on items go through an X-ray system.
Since different materials absorb X-rays at
different levels, the image on the monitor
lets the machine operator see distinct items
inside a piece of luggage bag. Based on
the range of energy that passes through the
object, items appear as different colours on
the display monitor, representing one of three
main categories: organic, inorganic and metal.
Machine operators are trained to look for a
broad range of suspicious items.
■ Security (body) scanners are in use at
Heathrow and are designed to ultimately
reduce the need for metal detectors and full
body-searches at airports. They make use
of low-power millimetre wave technology,
with the energy projected by the scanner
being 100,000 times less than a cell phone
transmission. All of Heathrow's scanners are
fitted with Automatic Threat Recognition (ATR)
software. The image produced is a generic
stick-like figure, with marks showing where
the scanner has detected concealed items.
Passengers are able to see this image as they

LEFT AND BELOW
Use is made of explosives and narcotics detection equipment – a simple wipe with a swab over checked or carry-on luggage is all that is necessary to collect a sample. In just eight seconds the colour-coded display presents results to the operator — red for a detection and green for the 'all-clear'. If a contraband substance is detected, the specific name is identified on the display. *(HAL)*

exit the scanner. No images are saved or retrievable at a later date.

■ In February 2010 the government announced the initial deployment of security scanners at Heathrow. This was part of a package of measures announced as a direct response to the attempted attack on Northwest flight 253 from Schiphol Airport in Amsterdam to Detroit on Christmas Day 2009, when a passenger tried unsuccessfully to detonate plastic explosives sewn into his underwear. The introduction of full-body scanners and other technology is a significant step towards a more robust defence against a backdrop of changing and unpredictable threats. The airport is working to make people more comfortable with their use.

■ Explosive detection is a non-destructive inspection process to determine whether something in a passenger's luggage contains explosive material. Most explosives are not water soluble, and it is very hard to get rid of traces on the hands even after washing with water and soap. Explosive traces can be found on undisturbed objects even months after the actual explosive has been removed. Several types of machines have been developed to detect trace signatures.

■ 'ePassport' Gates in operation at Heathrow are based on facial recognition biometric

technology. Biometrics essentially means checking fingerprints, retinal scans, and facial patterns using complex computer systems to determine if someone is who they say they are – or if they match a list of people the government has determined may be a potential threat.

■ Security staff at Heathrow are specially trained in the use of behavioural detection techniques to single out passengers behaving suspiciously, who are then subject to additional security measures or referred to the police or border authorities.

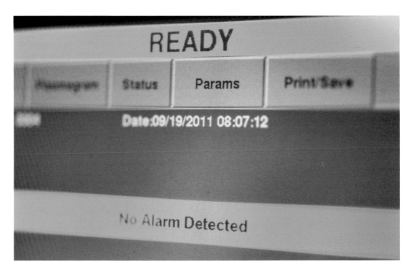

ABOVE Hold baggage is subject to rigorous security checks at Heathrow. *(Author)*

boarding a flight at Heathrow are electronically matched with their baggage in the belly of the aircraft. As is sometimes the case, in the event a passenger does not board the flight, his or her luggage must be off-loaded before the aircraft is allowed to depart.

The challenge of cargo security is addressed in Chapter 10.

Hold baggage

Every day more than 200,000 bags are put through Heathrow's security system. The screening of hand baggage is a challenging process requiring a high commitment of motivated manpower and advanced resources. Enhanced X-ray images are used to analyse the composition and contents of each bag before it is allowed on an aircraft, but the final decision still relies on an operator to make the detection decision.

In 1997 the United Kingdom became the first country in the world to adopt the practice of screening every piece of hold baggage. Checked baggage going in the hold of any aircraft is typically put through a CT scanner capable of calculating the mass and density of individual objects. Once scanned, hold luggage is held securely airside until it is safely onboard an aircraft.

It's reassuring to know that all passengers

The future of airport security

Today's passenger security screening process works – but at great cost to airports, authorities, the airline industry, and to passengers themselves. Globally, more than £4.7 billion is spent each year to keep aviation secure, and given the predicted growth in air travel – and continuously evolving security threats – today's model is not sustainable for the long term.

'Today's checkpoint was designed four decades ago to stop hijackers carrying metal weapons,' says Giovanni Bisignani, IATA's Director General and CEO. 'Since then, we have grafted on more complex procedures to meet emerging threats. We are more secure, but it is time to rethink everything. We need a process that responds to today's threat. It must amalgamate intelligence based on passenger information and new technology. That

AIRCRAFT SPOTTING AT HEATHROW

Aircraft spotting is a popular pastime involving the observation and logging of the registration numbers of aircraft, and Heathrow is at the top of many spotters' lists, given the volume and diversity of inbound aircraft. There are a number of regular viewing points, and Heathrow's security team and police have developed a programme to get enthusiasts involved where they can.

(Waldo van der Waal)

'Heathrow Airport Watch', set up by SO18 Aviation Security in November 2008, recognises the key part aviation enthusiasts have to play in keeping airports safe and acknowledges that they are a valued part of the airport community. The scheme provides members with a specially designed identity card, lanyard and cardholder which is to be worn at all times while enthusiasts are engaged in their hobby at Heathrow. It allows genuine enthusiasts to be easily identified by police and security teams at the airport. Regular visitors to Heathrow are likely to notice something out of the ordinary on or around the airport, and it is therefore hoped that the initiative will encourage enthusiasts to contact the police if they do see anything suspicious, while allowing the police and security teams to easily identify those who are genuinely enjoying their pastime at the airport.

Since the closure of the roof of the Queen's Building next to Heathrow's Terminal 2 over 20 years ago, the most popular viewing spot is undoubtedly on Myrtle Avenue on the south-east corner of the airport. It looks like any typical suburban street, but at the end of the road there is a patch of grass where spotters congregate daily to observe arrivals on runway 27L.

means moving from a system that looks for bad objects, to one that can find bad people.'

To this end, IATA is pursuing a 'Checkpoint of the Future' project, designed to find ways of enhancing security while reducing queues and intrusive searches at airports by using intelligence-driven risk-based measures. The holy grail is an uninterrupted journey from kerb to aircraft door, where passengers proceed through the security checkpoint with minimal need to divest, where security resources are allocated based on risk, and where airport amenities can be maximised.

IATA's Checkpoint of the Future would end the one-size-fits-all concept for security. Passengers approaching the checkpoint will experience a screening system tailored to them. The determination will be based on a biometric identifier in the passport or other travel document that triggers the results of a risk assessment conducted before the passenger arrives at the airport.

The main concepts of the Checkpoint are strengthened security by focusing resources where the risk is greatest, supporting this risk-based approach by integrating passenger information into the checkpoint process, and maximising throughput for the vast majority of travellers who are deemed to be low risk with no compromise to security.

Linked to this are improvements in screening technology that will allow passengers to walk through the checkpoint without having to remove clothes or unpack their belongings. Moreover, it is envisioned that the security process could be combined with outbound customs and immigration procedures, further streamlining the passenger experience.

Ultimately, Heathrow relies on third party technology companies to research and test new equipment to interrogate risks. These are inevitably expensive and time-consuming projects. At the same time, developers will want long-term contracts and a commitment to purchase their kit in volume, and all the while legislation might change, so the security challenge is something of a balancing act for the airport.

That said, developments like the Checkpoint of the Future will help Heathrow to expand an increasing trend towards multi-level surveillance and intelligence systems which are designed to monitor the entire airport, providing complete real-time situational awareness, thereby allowing the security team to make the best decision on how to neutralise any threat.

HEATHROW PEOPLE: CHRIS KERR

Acting Sergeant SO18 Armed Response Team

As an armed response officer, my team and I are tasked with providing an enhanced level of security across the airport. A morning briefing session, provided by our intelligence unit on both international and domestic terrorism threats, is generally the first order of the day. The session also provides new information on wanted suspects for whom we need to be on the lookout should they attempt to pass through the airport.

Preparing for a day in the field requires us to head to the armoury to book out and make ready our weapons. These typically consist of pistols, tasers and more powerful assault rifles. We also prepare our vehicle with the gear we need for the day, including a full first aid kit, a 'method of entry' kit and ballistic shields.

Typical tasks include a roving patrol of the airport's perimeter. Our vehicles carry a sophisticated automatic number plate recognition system we use to identify suspicious vehicles and our patrols are extended to include several of the airport's arterial roads, including the M4 and M25 spur roads as well as the A4.

Another task – which goes by the unusual acronym of HIPPO, or High Intensity Police Patrol Option – sees our team and another providing a high-visibility presence in one of Heathrow's terminals. There is no particular increase in the threat level, but our presence provides passengers with an enhanced level of reassurance. It also allows us to monitor the check-in process for a flight of interest to us. The airport and the airlines employ stringent security procedures, both on the ground and in the air, and our unit is tasked with providing an additional show of strength.

We spend a lot of time planning and conducting 'dry runs'. One such example is when we provide security cover for a foreign president due to land at Heathrow – it's often the case that high-level VIPs are afforded protection by the Metropolitan Police. I rehearse counter-sniper cover with other rifle officers located in various positions, while other armed response officers rehearse vehicle tactics using one of our unit's armoured Jankel vehicles, used in the event of a contingency. Everything is meticulously planned and carefully rehearsed to ensure nothing is left to chance.

We also get called in to deal with aircraft-related incidents. Our team was recently tasked to deal with a disruptive and aggressive passenger. Fortunately by the time the aircraft had landed our presence was enough to subdue the passenger and we simply had to escort him off the aircraft and hand him over to another police unit. On this occasion it looks to have been a case of nothing more than one too many from the drinks trolley.

CHAPTER 7

Airfield operations

OPPOSITE Airfield operations can become a significant challenge in adverse weather. *(HAL)*

Introduction

Heathrow's airside operation is a dynamic, fast-paced world that never sleeps. Daytime is dominated by a continuous stream of arriving and departing aircraft, the lifeblood of the airport moving people and cargo to and from the far corners of the planet. Late at night and through the early morning, an army of maintenance teams set to work to repair, install, and upgrade facilities and infrastructure to make certain that the airfield is ready for the next day's flying schedule to commence.

Ensuring the safe and efficient operation of the airfield is a complex task that requires Heathrow to comply with international rules established by the International Civil Aviation Organization (ICAO), the European Aviation Safety Agency (EASA), and the national regulator the Civil Aviation Authority (CAA). In addition there is compliance with strict health and safety regulation and the company's own organisational policies and procedures.

Like all airports in the UK used for transporting passengers or for flying training, Heathrow is licensed to operate by the CAA, and in the issuing of this licence the CAA must be satisfied that the physical conditions of the airport's operational areas and environs are acceptable, and that the scale of equipment and facilities provided are adequate for the flying activities which are expected to take place.

At the centre of a triangle maintaining the requirements of this licence and the efficiency of the airfield's daily operations is the Duty Manager Airside (DMA) and the Airside Safety Department (ASD). They work hand-in-hand and it is their knowledge, processes and experience, combined with an exceptionally high standard of training and a commitment to safety, which ensures Heathrow can operate at nearly full capacity every day.

The three corners of the triangle are the airlines, air traffic control and the variable factor that is the weather, with the DMA and ASD maintaining a strategic overview of the airport and pulling the relevant strings to maintain efficiency levels.

This chapter explores how Heathrow's airside operations team ensures the functionality of the facilities and infrastructure, while coping with the seemingly constant flow of aircraft, vehicles and passengers. Think of them, perhaps, as the station master operating a large train set with points, lights and tunnels that all need to be kept running if the train set is to operate as per the instructions that came with the box.

Airfield layout

The key areas managed by the Airside Safety Department are principally the runways and taxiways (collectively referred to as the manoeuvring area) and parking stands (sometimes referred to as the apron area and is equipped with jetties or airbridges to allow passengers to embark and disembark, and provide, amongst other services, electrical power to allow aircraft systems such as the cockpit instruments and cabin interior lighting to be used without having to run noisy, polluting, onboard power units). Stands are also the areas where ground-handling companies carry out the aircraft 'turnround', including activities such as cargo handling, refuelling, catering and first-line maintenance (a little like checking the oil and water on your car).

The airside operations team

Situated in the centre of the airfield is the Airside Safety Department (ASD), whose role it is to ensure the safe and efficient operation of the airfield, including the aircraft, vehicles and people. Parked outside are several of the team's distinctive yellow vehicles that are used to patrol the expanse of the airfield. This facility and its personnel, many of whom have worked at the airport in varying roles for more than 20 years, are the 'eyes and ears' of the airside operation. They are absolutely integral to safe and efficient operations, allowing Heathrow to continue operating at capacity. The team are at work in all weather conditions and in all operating circumstances and have an extensive remit of responsibility that includes:

- Ensuring that the runways, taxiways and apron areas are fully operational and in a safe state to use.
- Monitoring the airfield lights, signs and markings to ensure compliancy with the airport's operating licence conditions.
- Ensuring that vehicles moving around the complicated system of roads and tunnels are moving freely and complying with airside driving regulations.
- Maintaining an environment free from foreign object debris (FOD) that could damage aircraft.
- Managing the habitat to ensure that bird activity is minimised, thereby reducing the risk of bird-strikes.

- Marshalling aircraft on to stands if automated stand entry guidance (SEG) systems are not operational.
- Routine surface inspections across the whole movement area to identify damaged infrastructure and FOD.

Leading this challenging effort is the Duty Manager Airside (DMA), who is responsible for delivering the operational effectiveness of the aerodrome and also has the responsibility out of hours for holding the aerodrome licence. Nine DMAs work shifts to provide full-time cover at the airport on a 24/7 basis, and assisting them in this task are:

- The Duty Field Manager (DFM) – responsible for the first-line man-management and day-to-day running of the ASD, and reporting to the DMA that the airfield is functioning and compliant.
- The Manoeuvring Area Unit (MAU) – two teams on the airfield are responsible for inspection and patrol duties, including the runways, all taxiways and grass areas. They are also responsible for managing any bird hazards, supporting operational procedures during adverse weather and providing a front-line response during emergencies. In addition they take responsibility for control of maintenance work in progress on the airfield and provide

a central communication role by being the on-ground eyes and ears for ATC.

- The Apron Area Unit (AAU) – a number of teams on the apron area carry out inspection and patrol duties on aircraft stands, apron areas, airside roads and footpaths. They also respond to all marshalling and leader duty requests to ensure a safe and efficient operating environment for aircraft and persons on and around the apron area. The work also includes monitoring the turnround of aircraft for safety.
- The Stand Allocation Unit (SAU) – this team is responsible for the allocation of stands for inbound flights to make the best use of the available space. Not all stands and aircraft are the same size, and this makes the allocation process all the more challenging. The SAU is also responsible for the publication, through electronic means, of the arrival times and departure information of aircraft using Heathrow.
- The Senior Operations Controller (based in the Stand Allocation Unit) – the SOC manages the day-to-day operation of the stand allocation team to maximise available stand capacity and manage any impact to the flow of aircraft.

One of the key tools used by the DMA to ensure that he and his team deliver a safe and efficient airside operation is the Heathrow Operational Efficiency Cell (HOEC). The DMA chairs the HOEC (based in the ATC tower), and members take part in a conference call four times each day. HOEC is made up of a representative of NATS in Swanwick, Terminal Control (London) and a dedicated representative from Heathrow's Met Office, together with representatives from major airline carriers. Other airlines also have the opportunity to dial into the conference call. These calls offer the opportunity to discuss the 'state of play' on the airfield, to confirm that all relevant systems are working, to advise of any operational conditions such as the closure of taxiways etc and other notable information that might impact operational efficiency. HOEC also assesses arrival demand and any potential delays.

Airfield inspections

Regular airfield inspections form a vital part of the ASD's daily activities. The inspection process is designed to enable the highest degree of safety to be maintained for aircraft operations and for the health and safety of airside workers. The inspection process is stipulated as part of the ICAO licensing requirements for the airport and involves the close and careful visual observation of airside areas.

To facilitate the inspection process, Heathrow's Movement Area is divided into 32 distinct zones, including runways, taxiways, grass areas, stands, and apron areas. Amongst many other things, inspections look for surface defects, lighting failures, driver behaviour and spillages. All areas outside the airfield perimeter where hazards may form risks to aviation safety are also inspected regularly.

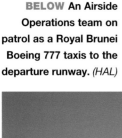

BELOW An Airside Operations team on patrol as a Royal Brunei Boeing 777 taxis to the departure runway. *(HAL)*

Inspections are broken down into three categories:

- Level 1 – routine daily inspections of the movement area.
- Level 2 – detailed airside inspections every 32 days.
- Level 3 – carried out by senior management accompanied by a DMA and engineering team on a quarterly basis.

Any issues or faults that are detected during the inspection process are registered with the maintenance team control and noted in a daily inspection log so that they can be dealt with appropriately. During runway inspections the team on duty are trained to be able to assess the severity of any particular situation, and will not hesitate to close a runway for safety reasons.

Dealing with adverse weather

The onset of adverse or extreme weather presents a new set of challenges to the airport's operational and safety teams. Low visibility, strong winds, thunderstorms, floods, frost, ice and snow are just some of the natural events which Heathrow has to contend with on a fairly regular basis.

Heathrow's runways are equipped with an Instrument Landing System (ILS), for low-visibility operations, allowing suitably equipped aircraft with trained crews to continue operations with meteorological visibility down to 50m and the cloud at ground level. The ILS consists primarily of two parts: the localiser (the orange 'toast rack' on poles at the very end of each runway) and the glide path or glide slope (a tall mast with three horizontal antennae offset to the left-hand side of a landing runway). Both pieces of equipment must be protected from interference, as their transmissions are 'listened to' by a landing aircraft; any break in signal could mean that the aircraft would have to divert off its approach and 'go around'. In adverse weather, the spacing between aircraft is also increased to allow an additional safety margin to landing aircraft clearing the runway.

Strong winds require the ASD to conduct additional inspections, particularly across the Movement Area, around any work in progress and on the runways. Extra time is spent checking the aprons, as air bridges in particular can be affected by high winds.

Thunderstorms bring the risk of standing water and flooding, while frost and ice require Movement Areas to be treated with de-icing liquid.

Located airside on the eastern side of the airfield is Heathrow's 'Snowbase'. It is used for the storage of snow-clearance vehicles and other associated heavy-duty winter operations equipment and fluids. The snow season runs from December to March and during this time the DMA is responsible for implementing a 'Snow Plan' that sets out snow-clearance operations for the whole Movement Area. Snow has to be cleared from runways and taxiways and ice and slush has to be dealt with on the ramp areas.

Given the implications of snowfall on airport operations, specific Met Office aviation forecasts for Heathrow are continually monitored to assess potential wintery events as far in advance as

ABOVE During their safety inspections, the Airside Safety teams constantly check to ensure there is nothing on the tarmac that could damage the tyres of an aircraft. *(Waldo van der Waal)*

its state of readiness and issues relevant 'Snow Alerts' to bring people and equipment to readiness.

In 2010 Heathrow was criticised when dealing with the impact of an exceptionally heavy snowfall that closed the airport. In mid-December of that year, the tenth coldest month for 100 years, heavy snowfall caused the closure of the entire airport, creating one of the largest incidents that Heathrow has ever experienced. More than 4,000 flights were cancelled over five days, causing significant impact to airline schedules globally. At the same time some 9,500 passengers spent the night at Heathrow following the initial snowfall. The problems were caused not only by snow on the runways, but also by snow and ice on the parking stands that were all occupied by aircraft.

Despite forecasts having predicted heavy snowfalls days before, the most significant feature of the snow event on 18 December was the rate at which snow fell, with nearly 7cm falling within the hour prior to midday.

Following a detailed investigation the airport has set about implementing a comprehensive Winter Resilience Programme, including:

- A new fleet of equipment used for the clearance and disposal of snow.
- Increasing the total number of staff available to clear snow.
- New agreements with airlines and ground handlers on jointly working to clear aircraft stands.
- Working closely with the airlines to better assess aircraft de-icing capacity. At Heathrow,

ABOVE Adverse weather can have a dramatic effect on efficiency levels as snow clearing is a time- and labour-intensive operation. (HAL)

possible. Once identified, a pre-programmed chain of actions is put in place to ensure that the airside operational teams have the support they may need, depending on the forecast conditions. Based on regular forecasting, the airport ramps up

RIGHT Clearing teams work in unison to clear the apron area on a remote stand around this Qantas Airbus A380. (HAL)

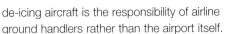

de-icing aircraft is the responsibility of airline ground handlers rather than the airport itself.

■ The development of new snow-disposal plans – in other words, where to put all the snow that has fallen once it has been cleared. A snowfall of 10cm would produce 48,000m³ of snow to be disposed of. This is equivalent to about 4,000 average lorry loads. In addition, any snow that is contaminated by de-icer has to be disposed of carefully and drained through the airport's pollution control system. Short-term storage is available on the airfield, but once these have been filled snow has to be transported off-site.

■ Appointing a full-time Winter Operations Manager responsible for the airport's winter readiness.

The climate and geography around Heathrow dictates that fog is a relatively common phenomenon, particularly in the early mornings, when, unfortunately, the airport has one of its busiest periods. Fog reduces visibility, and just as vehicle drivers have to leave more space between them and the car in front, the situation is no different with aircraft. With more space between arriving and departing aircraft, not as many can land or take-off each hour (a reduced 'flow rate').

The vast majority of major international airports have spare runway capacity, so that while spacing is increased and delays build there is sufficient capacity in the schedule to recover from this later on. Heathrow is unusual in that it operates at more than 99% capacity, with an aircraft landing or taking off every 45 seconds. Having to space

ABOVE LEFT Large volumes of snow can be moved at one time. (HAL)

ABOVE A snow tractor clears an access road for use by apron traffic. (HAL)

LEFT Heathrow's winter resilience investment includes in snow clearing vehicles such as these with cutter and blower attachments. (HAL)

clearance. Aircraft are also more widely spaced when manoeuvring or taxiing at the airport.

Meteorology

Weather conditions can change in what seems like an instant, and this poses many challenges for an airport, particularly so at Heathrow given the capacity challenges it faces. Therefore accurate and reliable weather information is a mandatory part of Heathrow's day-to-day operations.

To this end, the Met Office has forecasters 'embedded' at Heathrow. 'The on-site forecaster's advice is immediate,' says Dan Boon, currently stationed at Heathrow Airport's Operational Efficiency Cell (HOEC). 'Having continuous, on-site support provided by an embedded forecaster provides direct access to dedicated meteorological information and advice.' Information provided at Heathrow comes from the central Met Office in Exeter.

The Met Office is the UK's National Weather Service and uses more than ten million weather observations a day, an advanced atmospheric model and a high-performance supercomputer to create 3,000 tailored daily forecasts.

'Having our own dedicated forecaster means he has good local knowledge, understands the demands of the airport and can proactively look at Heathrow-specific issues to help maximise efficiency,' says DMA Mark Sandford.

Additional local information is provided by on-site technology, including a Semi-Automated Met Observation System (SAMOS) located in the 27L and 09L glide slope areas that provide routine meteorological observations for use by ATC. These comprise a number of sensors, measuring barometric pressure, cloud base height, visibility, temperature and humidity. The system measures, processes, records, and displays this information to users in the control tower. Runway temperature sensors, weather stations across the airfield and a cloud base measuring system also enhance the available meteorological information.

flights out more during fog inevitably causes delays and cancellations because there is simply no room to accommodate the delayed flights. Generally speaking, arriving aircraft are spaced around three miles apart. In low visibility that figure increases to six miles.

Although aircraft are equipped with technology that enables them to fly through fog, the issue rests with safe manoeuvring on the ground. In Low Visibility Procedures (LVPs) the preceding aircraft has to be allowed to land and clear the runway before the following aircraft is given landing

2010 VOLCANIC ASH CLOUD

In an unprecedented event, flights from across Europe were suspended for five days in mid-April 2010 due to the risk of jet engines being damaged by volcanic ash in the upper atmosphere caused by the eruption of the Eyjafjallajökull volcano in Iceland.

The controlled airspace of many European countries was closed, resulting in the largest air-traffic shutdown since the Second World War. The closures caused millions of passengers to be stranded not only in Europe but across the world. After an initial uninterrupted shutdown over much of northern Europe from 15 to 23 April, airspace was closed intermittently in different parts of Europe during the following weeks, as the path of the ash cloud was tracked. Criticism was levelled at aviation authorities, but no risks could be taken as the circumstances were largely unknown, and no data or previous experience existed in relation to dealing with an ash cloud.

Heathrow staff described the airport as a 'ghost town' during this period, with not an aircraft sound to be heard. Once the all-clear was given the airport began to restart operations, but this was no easy task, since many crews and aircraft that should have been at Heathrow were scattered across the globe.

Emergencies

Major emergencies at Heathrow are rare. However, airside teams are fully trained and equipped to deal with them should they occur. The majority of emergency calls are initiated by ATC, as they are in constant radio communication with aircraft approaching or departing from

Heathrow. Here are some of the situations the airport needs to be prepared for:

- Aircraft Accident – this is declared when an aircraft accident has occurred on the airport.
- Aircraft Accident (Off Airport) – declared when an aircraft accident has occurred outside the airfield perimeter fence.
- Aircraft Accident Imminent – declared when an aircraft accident is considered inevitable, either on or near to the airport.
- Aircraft Ground Incident – declared for an aircraft on the ground involved in an incident that may endanger aircraft or passengers. This may include a bomb warning.
- Full Emergency – declared for an aircraft in flight with a problem that could result in an accident. This may include hijacks or bomb warnings.
- Local Standby – declared when an aircraft has a minor defect that would normally be no problem when landing, when an aircraft requires external assistance or inspection or when an aircraft is the subject of a bomb threat. It might also apply when weather conditions are causing aircraft operational difficulties or a medical evacuation flight is arriving, departing or refuelling.
- Domestic Fire/Special Services Procedure – this typically applies to a fire that does not involve an aircraft or when the Fire Service is required to extricate a trapped person, render first aid or make safe a fuel, cargo, or substance spillage.
- Hijack – when any person on an aircraft, by the use of force or threat of any kind, seizes the aircraft or exercises control of it.
- Threat to Aircraft – if a bomb or suspect article is found on an airborne aircraft then a Full Emergency will be declared.
- Threat to Aircraft (Ground) – declared when a bomb threat has been made against airport infrastructure, property, or a location beneath the flight path.
- Act of Aggression – an act of terrorism, armed attack, bomb attack or hostage situation (other than hijack) taking place within, or adjacent to, the airport boundary.

Once an emergency has been declared, the various services spring into action. Heathrow's staff must call upon their training, experience, common sense and a degree of flexibility in order to deal with diverse situations. In the event of an

BRITISH AIRWAYS FLIGHT BA38 (BEIJING TO LONDON)

British Airways Flight 38 landed short of runway 27L at Heathrow on 17 January 2008 after a 5,000-mile (8,100km) flight from Beijing.

After an extensive investigation, ice crystals in the fuel were found to be the cause of the accident, clogging the fuel-oil heat exchanger (FOHE) of each engine. The accumulation of ice had no effect on the flight until the final stages of the approach into Heathrow, when increased fuel flow and higher temperatures suddenly released the ice back into the fuel. This formed a slush of soft ice which flowed forward until it reached the FOHE, where it froze once again, causing a restriction in the flow of fuel to the engines.

The aircraft passed above traffic on the airport's Southern Perimeter road and landed on the grass approximately 900ft (270m) short of runway 27L. The captain declared an emergency, calling 'Mayday' to the control tower a few seconds before landing. During the impact and short ground roll, the nose gear collapsed, the right main gear separated from the aircraft penetrating the central fuel tank and cabin space, and the left main gear was pushed up through the wing. The aircraft came to rest on the threshold markings at the start of the runway. A significant amount of fuel leaked, but there was no fire. In less than two minutes the Airport Fire Service and Airside Safety team put their full emergency plans into action to ensure the risk of any fire was controlled and that all passengers were evacuated and cordoned safely. Remarkably, there were very few injuries: four crew members and eight passengers received minor injuries while one passenger suffered from concussion and a broken leg.

The emergency teams then started to deal with isolating the incident from the rest of Heathrow's operation and returning the affected runway to service as quickly as possible. This involved many carefully planned and orchestrated tasks as well as recovery of the aircraft.

The 150-tonne aircraft was the first Boeing 777-200ER to be written off in the model's history. The problem was identified as specific to the Rolls-Royce FOHE, and the engine manufacturer subsequently developed a modification for its Trent 800 engine, with all affected aircraft retro-fitted with the modification before 1 January 2011.

BRITISH AIRWAYS FLIGHT BA762 (LONDON TO OSLO)

Another good example of the type of incident the ASD and emergency services at Heathrow train to deal with occurred on 24 May 2013, when a BA Airbus 319 carrying 75 passengers to Oslo was forced to return to the airport and make an emergency landing after black smoke and flames were seen coming from an engine.

Having taken off from Heathrow in a westerly direction from the southern runway (27L), the pilots were advised by ATC that their aircraft had left debris on the runway. Alerted by ATC and by Heathrow's specialist FOD radar, airside operations teams were instantly dispatched to the runway to assess the situation. The runway was immediately closed so that evidence could be logged and authority sought from the AAIB to clear it.

Moments later on the flight, the cabin crew – alerted by passengers – told the pilots that panels were missing from the engines, although the aircrew were already aware of a problem. The panels, or fan cowl doors – each weighing some 40kg – had detached and caused substantial damage to one of the aircraft's hydraulic systems, and a significant fuel leak had started. The crew declared an emergency and prepared to return to Heathrow. During the approach the starboard (right) engine caught fire. The engine was shut down using built-in extinguishers, but the fire was not put out until the aircraft landed.

What followed was a carefully executed and co-ordinated response from ATC, the ASD and emergency response teams.

The aircraft was given preference to land by ATC and immediately headed to Heathrow's northern runway (27R). On landing, the airport Fire Service was immediately on-scene tackling the fire, and, when safe to do so, passengers evacuated the aircraft on the runway using the emergency chutes. Fortunately, only three people had to be treated for minor injuries.

Both runways were closed for 30 minutes, as the majority of the airport's Fire Service and airside operations teams were dispatched to deal with the incident.

Given Heathrow's capacity constraints the runway closures had a significant knock-on effect, with 192 flights cancelled and 22 aircraft diverted to other airports.

Once the scene was secure and the emergency over, the aircraft was towed to British Airways' maintenance base where it could be inspected by the AAIB and repaired by the airline before being returned to service.

A preliminary AAIB concluded that a maintenance error resulted in engine parts detaching from the aircraft. At least one of the cowls struck the fuselage, and also caused minor damage to the wing, landing gear and a fuel pipe. Both of the engines were exposed, and the right-hand one caught fire and was shut down by the pilots.

Incidents such as this are extremely rare. Nevertheless, the ASD and emergency teams train constantly to ensure that they are prepared to handle such situations in a safe and professional manner.

aircraft accident, the police have overall control and responsibility for the incident; however, they depend on the Airport Fire Service to initially tackle the situation.

Invariably members of the ASD are first on the scene and take responsibility for collecting survivors together and distributing waterproof clothing and blankets as necessary. Care is taken not to place injured people in operational vehicles, as this can cause further injury or result in the vehicle being taken out of service at a critical time.

Inner and outer cordons are established by the Fire Service. The former extends from the fire appliances forward to the incident while the latter can be at much greater distances away depending on the operational situation. Normally only personnel with full protective equipment, including breathing apparatus, are permitted to enter the inner cordon.

No debris or parts can be moved from the scene, as these will certainly form part of the investigation carried out by the Air Accidents Investigation Branch (AAIB). As and when the AAIB remove parts from the scene, the exact location of each item is recorded and photographed.

There are up to four Reception Centres at predefined locations across the airfield, which are used as a central point for survivors, friends and relatives and as a reunion centre. Representatives from the airline involved play a critical role in communicating information and dealing with families and the media.

In the event of a major aircraft emergency there are a number of key considerations for the airport to examine, including:

- Establishing the facts about the incident.
- Checking that those on board, whether fatalities, casualties or uninjured survivors, are being looked after appropriately.
- That next of kin are looked after and kept informed.
- Keeping the media informed.
- Dealing with the knock-on operational effects of the incident should the airport and/or its runways be closed for any period of time.

In any emergency the DMA is under pressure to return the operation to 'business as usual', so he or she must isolate any incident and seek to minimise its impact on the airport while still maintaining focus on dealing with the incident or emergency itself.

Fire, rescue and medical services

Fire and rescue services are a requirement at airports across the world. The number of firefighters and fire appliances (and their capacity) governs the size of aircraft that can be handled at an airport. This capability ranges from category 1 (lowest) through to 10. It comes as no surprise, then, that Heathrow is listed as a category 10 facility given that it caters for the world's biggest commercial aircraft and therefore requires extensive specialised rescue and firefighting cover.

By law, if Heathrow has no fire cover available it has to close its runways to passenger aircraft unless in an emergency, which is then at the discretion of the captain of the affected aircraft. These cases are extremely rare and would typically mean that the full complement of AFS resources is dealing with another major emergency elsewhere on the airfield.

The airport's main fire station is centrally located on the airfield, but to enable the Fire Service to meet the required response time of two minutes to all parts of the airfield a second station, called 'Fire East', is located adjacent to the Snow Base on the eastern side of the airport.

Heathrow's firefighters specialise in dealing with complex fires and rescues from aircraft, and a great deal of their daily routine is spent training and drilling for such eventualities. In support of this the Fire Service makes use of a dedicated training ground, also located on the eastern side of the airfield. Easily spotted by passengers arriving

at the airport from the east, and affectionately known by staff as the 'Green Giant', fire-training drills are focused around a full aircraft mock-up. This 'simulator' is used to replicate more than 35 different fire emergency situations inside the

ABOVE **Fire crews during a drill on the apron.** *(HAL)*

LEFT **Fire power – whenever an aircraft commander declares a potential emergency, the vehicles are brought to standby immediately.** *(Waldo van der Waal)*

BELOW **Heathrow's main fire station.** *(Author)*

aircraft, including the cabin, cockpit and galley, as well as external parts of the aircraft including the undercarriage, engine and wings.

'Heathrow has recently invested in a new range of major foam tender vehicles capable of carrying an enormous amount of water and foam concentrate, as well as other fire-extinguishing material, which is applied under massive pressure and volume at the scene of a fire,' says the head of Heathrow's Airport Fire Service (AFS), Les Freer. 'The vehicles have roof-mounted monitors [water cannon] that can project fire-extinguishing fluid over large distances, allowing an approaching fire engine to begin tackling flames before it has arrived at the scene.'

In addition the vehicles are each equipped with an infrared camera. Forward-looking infrared

LEFT The Heathrow fire service employs the use of a 42m aerial ladder platform. (Waldo van der Waal)

(FLIR) equipment assists the vehicle crew when responding to incidents in darkness and low visibility, as well as identifying 'hot spots' on aircraft, such as overheated brake assemblies, which pose a fire risk or may be dangerous to the fire crews.

To cope with a possible fire on the sizeable A380, the AFS is equipped with a 42m aerial ladder platform to get to the upper reaches of the aircraft, while some firefighting vehicles utilise a boom that extends to deliver foam to the aircraft's upper deck. Other equipment includes two command vehicles, a hose-layer and several auxiliary vehicles (such as a personnel carrier), along with a couple of reserve foam tenders.

In the same way that the airport has its own police and fire stations, so Heathrow has its own ambulance station too, manned by the London Ambulance Service, which is supported by first responders who operate in all terminals on mountain bikes.

Work in progress

With an airport as busy as Heathrow, the majority of airfield works can only be carried out during the night period when there are no flights.

On a site with as many developments and so much ongoing maintenance, including runway refurbishments, Heathrow is turned into what is sometimes described as 'one of Europe's largest construction sites' overnight, and then turned back again into an international airport in time for the next morning's first flights.

LEFT London Ambulance Service cycling paramedic team in Terminal 5. (HAL)

BELOW Even the smallest issues are quickly dealt with. In this instance a taxiway repair team undertakes some minor resurfacing work. (Author)

Heathrow's ASD and DMA have to know exactly what work is being carried out and how it might impact operations. To manage this requirement the airport operates a comprehensive permit system for contractors. With this in place, the ASD team safeguard those working airside from the continuing operation of the airport around them. They also safeguard the airport from those engaged in airside works and safeguard flight operations with continued vigilance of the surrounding areas and within the airfield when crane operations are under way.

Aprons and stands

An apron is a defined area intended to accommodate aircraft for the purpose of loading or unloading passengers, mail or cargo, refuelling, parking or maintenance. At Heathrow the apron areas are divided into individual stands and the road network.

Stands are constructed of concrete and/or block paving (asphalt cannot be used due to the problems of spillage contamination). Airside planners work to ensure that there is enough apron space to provide for the number and types of aircraft expected to use it. They also ensure there are adequate safety margins from obstructions, including other parked aircraft. The apron is designed with the aim of facilitating the movement of aircraft and avoiding difficult manoeuvres that might require undesirable use of excessive amounts of engine thrust, or impose abnormal stress on tyres.

Floodlighting is provided for all aprons intended for use at night. The floodlights are located to illuminate all areas of the apron without causing glare to pilots, ATC or ground staff. The challenge of stand management is dealt with in the chapter on airline operations.

Surface run-off management

Heathrow generates a large amount of run-off water when it rains, which must be managed appropriately before it leaves the airport. This is required to protect local watercourses from the effects of flooding and any contaminants present, particularly during the winter months when de-icing fluids are used at the airport to ensure the safety of aircraft.

Heathrow has invested in pollution control infrastructure to contain and treat surface water run-off. This includes reservoirs, reed beds and balancing ponds to store run-off, and advanced biological treatment facilities. The airport's surface water pollution control system enables it to hold on to water run-off when de-icing fluids have been used on aircraft and ground surfaces, and release the water into the environment when it meets standards set by the Environment Agency. Heathrow continues to look at different ways of recovering de-icing fluids and deploys vehicles to remove residual aircraft de-icing fluid from stands. The airport has an established process, equipment and personnel to deal with incidents by containing any harmful substance as close to the source as possible and before it reaches local watercourses.

Significant investment is planned over the next five years to further improve the airport's pollution control infrastructure.

Ground transportation system

If anyone ever needed just one way to demonstrate that Heathrow is indeed a small city it would be the fact that the airport has some 26,000 licensed drivers on site, doing everything from delivering baggage and carrying out inspections to ferrying passengers on buses from car parks and even towing aircraft to remote stands.

'We have belt-loaders transferring suitcases from transport dollies into aircraft holds, we have pushback tugs, so called scissor-lift trucks which hoist catering supplies to the passenger decks of waiting aircraft, maintenance vehicles and even sewage bowsers – or "honey trucks", as they're known – in action around the clock at Heathrow, so we keep a close eye out for safety breaches such as driving around with open doors or

speeding,' says Rachel Heydon, from the Airside Safety Practice Team.

With the airport's clear focus on safety, Heathrow's road and traffic system operates in much the same way as in everyday life, with specific road markings, speed cameras in key locations and even a points system that penalises careless drivers.

ABOVE An aircraft tug on the apron area, with Terminal 2 construction in the background. *(HAL)*

HEATHROW PEOPLE: SIMON NEWBOLD

Airside Operations Training Manager

As the Airside Operations Training Manager, my day-to-day role is focused on the vital area of training members of the Airside Safety Department (ASD) in the safe operation of the airport's runways and taxiways. This includes everything from dealing with bird hazards to the safe marshalling of aircraft. We're based in the centre of the airfield for a reason – our location allows us to respond quickly to any developing situation or emergency.

I spend a fair amount of time running training courses for the airport's airside safety teams, which emphasise the need to develop a strong sense of awareness, to look one step ahead and to adopt the approach that 'if it doesn't look right, it probably isn't'. Spotting something of concern isn't a skill that can easily be taught – it comes more from experience, so that the best form of education is not in the classroom, but out driving around the airfield. It's something I try to instil from an early age in our apprentices who come to Heathrow for their first taste of what life is like airside. I find it a real pleasure to work with youngsters who have a real thirst for knowledge. The challenges of airside safety are so varied that it's impossible to teach someone enough to cover every possible eventuality, but my aim is to give the team sufficient collective knowledge and the necessary tools to deal with pretty much anything that comes their way, be it an airside emergency or adverse weather.

Joining one of the recently qualified crews as they carry out one of the daily runway inspections looking for foreign object debris I get great satisfaction from seeing how well the team

perform their task. Just a year ago they were new to the airfield, nervous on the radio, and found it difficult to tell the difference between aircraft types. Now they are well established, confident, and carry with them a great deal of responsibility.

Despite all my time at the airport, I can honestly say I never fail to learn something new every day. Heathrow is constantly evolving; legislation is changing, and we need to adapt accordingly, and, at the same time, so is the airport's topography – all the time, which in itself presents new challenges.

CHAPTER 8

Airline operations

OPPOSITE A Virgin Atlantic Airbus A340-600 on a taxiway during a push manoeuvre with a tug. *(HAL)*

Introduction

The 84 different airlines and their passengers that use the airport are undoubtedly Heathrow's most important customers, and much of the daily operations are structured to meet their complex and time-sensitive needs.

Invariably the success of airline operations comprises a combination of the services provided by the airline itself, those of its handling agents on the ground and those provided directly by the airport. Whether it's the airport providing leader duties to the stand for a pilot unfamiliar with the airport's layout, the baggage team that ensures passengers are reunited with their luggage shortly after landing or the airline's operations team that takes responsibility for de-icing an aircraft in freezing conditions, one thing is absolutely clear – it's an all-round team effort that has to be choreographed to the minute, given the constraints at Heathrow.

This chapter explores how the airport, airlines and service agents work together to ensure that an aircraft can safely land, meet its turnround time schedules and safely depart. A missed slot could compound the delay and is both costly and disruptive to the airline and inconvenient for the passengers, making an on-time departure a critical factor for every airline.

STOCKING A BRITISH AIRWAYS 747-400

From toilet rolls to teaspoons, British Airways loads thousands of individual items on to each jumbo jet before it takes to the skies. With a combined weight of 6,120kg, the items have to be unloaded and reloaded before every take-off. On a typical BA 747-400 departing on a long-haul flight, the following items are loaded:

- 1,263 items of metal cutlery.
- 1,291 items of china crockery.
- 538 meal trays.
- 735 glasses.
- 650 paper cups.
- 34 metal teapots.
- 220 drinks stirrers.
- 500 coasters.
- 233 toothpicks.
- 2,000 ice cubes.
- 99 full bottles and 326 quarter bottles of wine.
- 700 small cans of fizzy drinks.
- 164 bags of nuts in Club World.
- 337 cushions and pillows.
- 337 blankets.
- 337 sets of headphones.
- 337 headrest covers.
- 435 air sickness bags.
- 58 toilet rolls.
- 40 extension seatbelts for children.
- 340 safety cards.
- 337 copies of High Life magazine.
- 40 'Skyflyer' packs for children.
- 5 first aid kits.

Rod Green, BA's Head of Global Supply Chain, says: 'It's a huge job getting a jumbo into the air, let alone a fleet of 52 every day. There are teams across the airline working together 365 days a year to ensure that all 27,260 items are delivered on time and to the right place to ensure our customers enjoy the very best travel experience. When we receive our new aircraft, the challenge will be even greater.' With the introduction of the A380 the number of items loaded on to the aircraft has increased by approximately 10,000 to cater for two full decks of customers.

Turnround services

This activity encompasses the many service requirements of aircraft from the time they arrive at one of Heathrow's gates to the time of the next departure.

The various ramp service functions are carried out by a number of different companies, and a collaborative effort is needed to ensure the turnround is carried out safely with speed, efficiency and accuracy. Aircraft don't make any money for their airline while sitting on the ground – they need to be in the air as much as possible, and to make this happen requires capable staff and a well-maintained fleet of vehicles and ground support equipment.

About 20 minutes before a flight is scheduled to arrive on stand, a small army assembles to prepare for the inbound flight. Leading the team is a Dispatcher (also known as a Turnround Manager), responsible for overseeing all aspects of the turnround, with a clear focus on safety, efficiency and the following of procedures for everything from refuelling and chocking to the placing of cones around the aircraft and the safety of ground staff on the apron.

The turnround plan changes for different aircraft types. When it comes to the turnround of a large aircraft such as an Airbus A380 or Boeing 747, up to 50 different people can be involved, so the Dispatcher has to closely supervise and coordinate the various third-party activities. The Dispatcher also carries out all the necessary preparation for the dispatch of the outbound flight and ensures

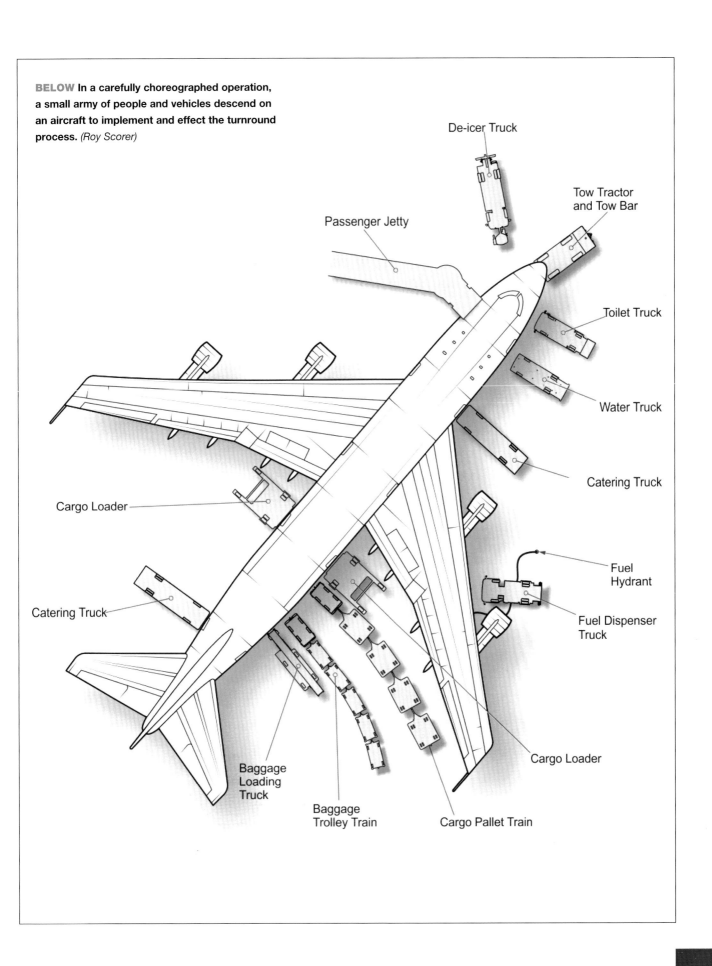

BELOW In a carefully choreographed operation, a small army of people and vehicles descend on an aircraft to implement and effect the turnround process. *(Roy Scorer)*

De-icer Truck

Tow Tractor and Tow Bar

Passenger Jetty

Toilet Truck

Water Truck

Catering Truck

Cargo Loader

Fuel Hydrant

Fuel Dispenser Truck

Catering Truck

Cargo Loader

Baggage Loading Truck

Baggage Trolley Train

Cargo Pallet Train

that the relevant flight documentation is distributed appropriately. He or she prepares and produces the relevant loading documents for a flight.

To put the scale of ground handling into perspective, a single long-haul Boeing 747 has over 40,000 items loaded on to it before it flies. All together these items weigh in excess of six metric tonnes and occupy a space of 60m³.

Ground handling encompasses several key functions, typically divided into ramp, catering and cabin services, and the efficient and seemingly flawless running of these operations are absolutely key to Heathrow's operating capacity.

Ramp services

Ramp services address the various service requirements of an aircraft between the time it arrives at a terminal gate and the time it departs on its next flight:

- Chocks – these are placed in front of and behind the nose wheel and the main undercarriage to prevent the aircraft from rolling forwards or backwards.
- Ground power – Provided so that the aircraft's engines need not be running to provide power when on the ground, reducing CO_2 emissions and noise. Also used to keep the cabin at a comfortable temperature and to power lights while the plane is boarding. In extreme weather conditions supplemental heating or air conditioning units are provided by separate ground-based units.
- Air bridge – connects the aircraft to the airport terminal and allows passengers to enter and disembark from the aircraft safely and smoothly.
- Baggage handling – carried out using belt loaders and baggage carts, which transfer passenger suitcases from the aircraft to the baggage system and ultimately on to the luggage carousel in the terminal.
- Fresh water supply – this involves using a water bowser to replenish the aircraft's potable (drinking) water supply for use by crew and passengers.
- Lavatory drainage – waste from the aircraft's lavatories is removed.
- Air-start units – these use compressed air at high pressure for starting engines on older aircraft.
- Air cargo handling – cargo on the aircraft is removed using cargo loaders and dollies.
- Refuelling – provided by a specialist third-party supplier which uses a dispenser vehicle to connect the aircraft to the fuel hydrant on the stand.

Catering services

In-flight catering makes use of a highly complex supply chain, with some airlines providing their own catering service while others outsource their requirements to third-party suppliers. Meals are prepared mostly on the ground in order to minimise the amount of preparation required in the air.

- Unloading – empty trolleys and catering waste are removed from the flight.

RIGHT A fully laden Emirates Airbus A380 from Dubai taxis as the turnround team makes final preparations for its arrival on stand at Terminal 3. (Waldo van der Waal)

LEFT Ground crew immediately undertake an inspection of the aircraft's four GP7200 turbofan engines. *(Waldo van der Waal)*

BELOW The Airbus A380 is served by multiple air bridges given its double-deck configuration and high volume of passengers. *(Author)*

BELOW LEFT Crew from ground handling specialists Dnata oversee the efficient unloading of passenger baggage. *(Author)*

BELOW RIGHT With the hold doors open the ground crew commence with unloading the cargo. From the stand it is trucked to the airline's freight distribution centre. *(Waldo van der Waal)*

- Loading – fresh food and drink for passengers and crew is loaded. Airline meals are typically delivered in trolleys from the catering company to the airport, where they go through security checks before being loaded on to an outbound aircraft.

Cabin services

These services are carried out predominantly to ensure passenger comfort.

- Cabin cleaning – cleaning of the passenger cabins and preparation of the cabin to the standard required by the airline. Sometimes the cleaning crew's job is made a little tougher than normal. For example, in June 2013 a Singapore Airlines flight en route to Heathrow hit some severe turbulence and suddenly dropped 65ft, leaving the cabin strewn with food, pillows, trays and cutlery. The cabin crew did what they could to clear up but the aircraft needed a 'full valet' before being returned to service.
- Replenishing consumables – includes items like soap, pillows, tissues, toiletry bags, first aid items, air sickness bags and blankets.
- Media – ensuring a supply of newspapers and in-flight magazines, as well as headsets.
- Duty free – goods sold on board have to be reconciled and restocked.

Leader duties

The Airside Safety Department (ASD) team is regularly called upon to provide various types of leader duties which involves guiding aircraft, works contractors and emergency services to various parts of the airfield. The role requires the team to have an excellent knowledge of airside topography, procedures and regulations.

- Static leaders are typically responsible for acting as the lookout at a works site, and to enable the works team safe access to and from the area. A member of the ASD team keeps the work team safe from potential aircraft hazards and ensures the site is kept clean and free of debris.
- Follow-me leader duties involve leading emergency services from a rendezvous point or security control post, leading a 'live' or towing aircraft, vehicle or equipment. The service is particularly key in low visibility or adverse weather conditions and can be used to assist pilots who are not familiar with the airfield layout, particularly one the size of Heathrow.

Each leader vehicle carries a 'follow-me' sign which is capable of displaying different messages – such as 'Follow Me', 'Slow' and 'Stop' – to the following vehicle or aircraft. When acting as an escort for a 'live' aircraft, ATC ask the pilot to follow the leader to the intended destination point. In summary, all leaders must have an ability to communicate with ATC to the highest standard using radio communications.

Marshalling duties

Sometimes humorously described as 'the men with the ping pong bats', airport marshallers use an internationally recognised set of signals to direct aircraft and are employed at Heathrow for a variety of reasons:

- When a stand entry guidance system is unserviceable, not provided, or in some instances not calibrated for the aircraft type.
- To assist a pilot who may be new to the airport.
- When extra guiding help is needed, such as work in progress in close proximity to where the aircraft must park.
- In low visibility when the stand entry guidance system is not clearly visible.

The pilot is ultimately responsible for the safety of his or her aircraft and is under no obligation to follow the signals of the marshaller, but pilots invariably put their trust in them because they are unable to see all the potential hazards affecting the aircraft and where to stop.

Marshallers must take care to always remain in view of the pilot, maintain adequate clearance between the aircraft and any obstacles and safeguard against any possible blast hazard. For all large aircraft they often work in tandem, with the 'main marshaller' providing signals to the pilot while the 'stop marshaller' gives the stop signal when the aircraft has reached its correct stopping position on the stand.

Stand management

The safe and efficient management of aircraft to and from the parking stands at Heathrow is essential to maintaining the airport's capacity levels. The airport currently has approximately 200 stands, the majority of which are air bridge served. Some remote stands use mobile steps and require

A member of the
Airside Operations
team guides an Air
Canada aircraft to its
parking position on
the stand using
fluorescent batons.
(Waldo van der Waal)

passengers and crew to be moved to and from the aircraft by bus.

Under direct responsibility of the Duty Manager Airside (DMA) is the Stand Allocation Unit (SAU), whose task is to provide the highest level of safety and customer service to airlines by allocating arriving aircraft to stands that are fully open, serviceable and unrestricted, maximising the use of air bridge and pier services.

Each evening the SAU team reviews the flight schedule for the following day in order to best allocate the stands available. Not all stands are the same size, and this makes the allocation process all the more challenging as aircraft must be sent to specific terminals depending on where the airline is based and whether they are international or domestic flights.

As Terminal 5 is used exclusively by British Airways and its sister airlines, in this terminal the airline manages its own stand allocations, with an overall check kept by the SAU team. For the remaining terminals, as well as the Royal Suite and cargo stands, all allocation is carried out by the SAU.

The closure or restriction of a stand can have a major impact on terminal operations, with a knock-on effect throughout the day. Reasons for closure include spillages, the failure of an air bridge, equipment infringing the stand, unplanned maintenance or problems due to adverse weather.

To draw up their daily plan the SAU makes use

of published scheduled time of arrival (STA) and scheduled time of departure (STD) information provided by the airport operators to Airport Coordination Limited (ACL), who administer the best utilisation of 'slots' on behalf of the airport. The definition of a slot is 'the scheduled time of arrival or departure available or allocated to an aircraft movement on a specified date at an airport'. It should be noted that ATC also issue 'slots', known as runway and airway slots – these are unrelated to airport slots but will be familiar to passengers, who may have heard the captain say, 'We've been given an earlier slot', or 'We've missed our slot and we're going to be on the ground for a while longer.'

Chapter 11 explains further how the ATC slot system is currently in the process of change to improve capacity-related issues across Europe by using what is called Airport Collaborative Decision Making (A-CDM).

Updated flight information is entered into the SAU's computer system, which in turn provides information to the internal Staff Information System (SIS) screens, public display boards and the Internet. The SAU is also responsible for the publication of all information on the SIS screens, including weather warnings or other information that is of an urgent operational nature.

Because Heathrow operates at near capacity, an absolute guarantee of immediate availability

of an operational stand on arrival is not always possible. In these rare circumstances the pilot is notified by ATC and the flight is published as a 'hold' until the stand is cleared. Similarly, last-minute stand changes happen from time to time, and ATC will again notify the pilot and re-route the aircraft to its new stand.

Airlines with flights on long turn-around times (such as South African Airways, which typically has flights arriving into Heathrow early in the morning and then departing again in the early evening) will be towed off the terminal gate-served stands and parked remotely. This allows Heathrow to maximise the use of prime stands and afford airlines the opportunity to carry out servicing and light maintenance during the aircraft's layover.

To maximise capacity, Heathrow also makes use of two flexible stand management arrangements. The first is entitled MARS (multi-aircraft ramp system), which offers flexibility for changing aircraft size by accommodating two small aircraft or one large aircraft parked on the same stand, for example two A321s or one B747. The second is MCA (multi-choice apron) – a defined area of apron accepting more complex combinations of aircraft than MARS, for example three smaller aircraft or two larger ones. The major advantage of aprons using MCA layouts is the flexibility provided to meet different aircraft mix requirements at different times.

ABOVE Good stand management is essential to maintaining the airport's capacity levels. *(HAL)*

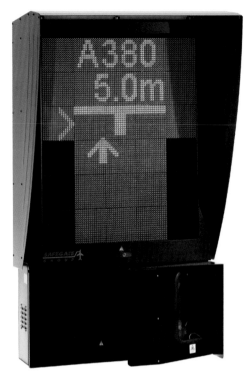

Stand entry guidance

Operational policy at Heathrow dictates that, as far as is reasonably possible, all arriving aircraft will 'self-park' on nose-in stands, using a stand entry guidance (SEG) system entitled 'Safedock'. Only certain stands are excluded from the nose-in principle due to their design, such as the Royal Suite and cargo stands.

The system is designed for minimum input from airport staff during operation and gives pilots all the information they need for safe and efficient docking procedures.

After vacating the runway, pilots follow ATC directions to their allocated stand. They then look out for the 'lead-in arrow', a yellow-painted ground mark with the stand number and an arrow adjacent to the taxiway centreline pointing towards the stand. The lead-in-arrows are either straight or curved to assist aircraft turning arcs.

An illuminated stand number indicator board (SNIB) at the head of the stand provides confirmation of the stand number. Each stand is provided with a yellow centreline for nose-wheel guidance.

Prior to an aircraft arriving on stand, the Dispatcher switches on the SEG following a complete physical stand inspection to ensure it is fully serviceable, free from any obstructions and clear.

The SEG is a single unit positioned at pilots' eye level at the head of the stand. It provides centreline and stopping guidance to accurately park the aircraft.

One of the advantages of the system is the level of accuracy it provides, allowing aircraft to be parked to within 4in (10cm) of the required point. This allows an air bridge to be pre-positioned, which in turn improves turn-around times and operational performance. Statistics show savings of up to 1.5 minutes per docking or 3.8 days of operational time saved annually per gate. Explained in a different way, the savings for a terminal with 30 Safedocks is equal to the capacity of an extra gate. The system consists of:

■ A single cabinet housing an LED display unit, laser scanner, power and control units mounted 4–8m above the ground.

- A display giving easy, clear and concise information and instructions for pilots during a docking procedure. The information includes: aircraft type, distance to stop, azimuth guidance (to show the required left or right steering input for the pilot) with instructions such as 'Stop', 'OK' and 'Slow down'.
- A laser rangefinder with a transmit and receive lens that scans a pre-defined docking area, and when an aircraft is detected and its approximate position confirmed it then measures the distance to the approaching aircraft and sends nose, wing, height and position data to a control unit for processing. It checks the graphical profile of the aircraft and compares it to corresponding parameters of the selected aircraft. If there is no mismatch between the parameters, the docking process continues.
- A control unit that processes the data in real time and submits the results for presentation on the LED display unit.
- Digital countdown clocks that appear as the aircraft nears its stopping point. Left and right adjustment arrows are also used if the aircraft is off the centreline.

After a successful docking the system reports 'OK' to the pilots.

Apron equipment

On each stand there is a set of equipment to assist with stand operations. At Heathrow this typically includes:

- FEGP (fixed electrical ground power) – provides electrical power to the aircraft while parked on a stand and helps to reduce noise and pollution by allowing the aircraft to shut down its auxiliary power unit as soon as possible.
- Pre-conditioned air – provides temperature control for the inside of aircraft cabins and also negates the need for aircraft engines to supply power.
- Emergency telephone – provided to allow staff

LEFT Apron equipment caters for all standard and emergency operations. *(Author)*

from the airport, airline or a third-party supplier to the aircraft to communicate with their offices if required.

■ Fuel hydrant emergency stop – for use if there is a major hydrant leak or fire in the vicinity.

■ Green bins – used for debris and general rubbish.

■ Red bins – used for pollutants, oils and lubricants.

■ Sawdust and grit bins – sawdust is used for minor spillages while grit is used in extreme winter conditions.

Aircraft refuelling

Heathrow is connected to the UK Oil Pipeline (UKOP) system, which pumps aviation fuel and other products produced from some of the country's largest refineries to various locations, particularly Fawley in Southampton, with two dedicated pipelines serving just Heathrow. On site it is held in large storage tanks located near the Fire Station and adjacent to the cargo apron on the south side of the airport. From these storage facilities the fuel is distributed through a series of ring mains that provide fuel to each stand via a hydrant pot.

The fuel installation is jointly owned and operated by individual suppliers. These organisations do not own any assets or fuel, but are involved solely with the transfer of fuel on to aircraft. Heathrow Hydrant Operating Company looks after the hydrant system itself.

Each flight and airline is contracted to a fuel company and the dispenser vehicle turns up at the stand when the in-bound flight arrives. It connects to the hydrant and then to the aircraft. The fuel in the hydrant is pressurised, so the vehicle has several functions – it not only filters and meters the fuel, it also depressurises it from 150psi to 50psi.

Quality control and contamination issues take up much of the refueller's time – the dispensers are fitted with filtering, monitoring and recording equipment as well as a hoist to enable operators to access aircraft with high wing-mounted couplings.

The advantage of hydrant fuelling over a tanker is the removal from an already congested stand of a large vehicle, and the speed and volume of fuel that can be transferred. For example, an A380 has

a capacity of 320,000 litres, whereas a tanker only has a capacity of 40,000 litres; therefore it would take some eight tankers to refuel the aircraft.

On all fuel-served stands there is an emergency stop button, designed to ensure the prompt curtailment of fuel flow in the event of an emergency or incident on the apron. Fuel spillages are treated carefully and may involve the Fire Service being called in and cleaning equipment requested on site.

During fuelling, air and vapour are displaced from an aircraft's fuel tanks. This potentially explosive vapour is expelled via vent points on the aircraft's wings.

In addition to providing fuel for the 650 daily outbound flights, there is also a requirement to fuel the 7,000 on-site vehicles which provide essential services, such as aircraft maintenance, buses to bring in staff, vehicles to push out and tow in planes, and machinery to load cargo, handle baggage, empty toilets, clear snow, de-ice wings and stock-up with catering supplies.

AIRCRAFT FUEL CAPACITIES			
All figures in litres			
Boeing		**Airbus**	
737	13,000	A319/320	24,000
747	226,000	A330 (200)	139,000
757	43,000	A340 (300)	140,000
767	46,000	A340 (500)	215,000
777	181,000	A350	138,000
787	126,000	A380	320,000

In-flight catering

The provision of in-flight catering for Heathrow's departing flights is a highly complex operation, and the way meals and equipment are transported and supplied has a close affinity to military-style logistics and distribution systems. The President of KLM Catering once famously described flight catering as '70% logistics and 30% cooking'.

When an aircraft arrives at Heathrow it is stripped of all its catering trolleys and equipment, including reusable crockery and cutlery. These are then returned to the supplier for washing and sterilisation.

For outbound flights, airline catering suppliers are required to orchestrate the assembly of meals according to specifications provided by

RIGHT Air bridge installations at the new Terminal 2. *(HAL)*

the airlines based on forecasts of passenger numbers on any given flight. Meals vary widely in quality and quantity across different airlines and classes of travel. Passengers' dietary requirements and restricted food products also need to be considered. Once the order volume is clear, a series of complex steps are then followed to produce trayed meals and non-food items ready for transportation to the aircraft. The service trolleys are packed in such a way as to make life as easy as possible for the flight crews.

Transportation to the aircraft is usually undertaken by specialist high-loader trucks that enable trolleys to be rolled on and off with ease. Once loaded, trolleys and other items are stowed on board, ensuring food safety and the security and safety of the crew, passengers and aircraft.

Air bridges

Heathrow's air bridges (sometimes called 'jetties') allow crew, staff and passengers to board and disembark aircraft. They provide a direct, efficient and safe link between the aircraft and the terminal. Prior to the introduction of air bridges passengers would normally board an aircraft by climbing a set of movable stairs or the aircraft's own stairs.

Air bridges not only provide protection from the weather, they also aid turnround times. Their use also improves safety at the airport, with less congestion on the apron by vehicles and passengers. An air bridge is also designed so that it can be used as a means of escape in the unlikely event of a fire aboard an attached aircraft.

Only appropriately trained personnel who hold a permit may operate an air bridge at Heathrow. The bridge is operated from a control console with buttons, a graphic display and a single multi-axis joystick.

RIGHT An air bridge connects an Emirates Airbus A380 to Terminal 3, providing safe passage for those disembarking from the aircraft. *(Author)*

Air bridges at Heathrow can be divided into two main categories:

- Rail drive bridges – these are capable of limited movement and are reliant on accurately parked aircraft.
- Apron drive bridges – the most versatile and widely used bridge in the world. They offer maximum apron flexibility, with an ability to operate over multiple centrelines serving the widest range of aircraft sizes. They are telescoping passageways that can be driven across the apron to the aircraft. They are more configurable than other bridge types and, importantly, offer flexibility regarding which aircraft is put at a specific gate.

All A380-served stands have three bridges that are named to correspond to the aircraft doors. This allows for faster boarding and disembarking, and allows separated entry to the aircraft for passengers in different classes.

Air bridges are fitted with an auto-levelling device that enables the air bridge mechanism to compensate for changes in the aircraft's height as its weight changes while being loaded or unloaded.

Weight and balance

A clear set of rules and methods are used in the aviation industry to ensure that an aircraft is not overloaded and that the weight distribution of the load being carried is balanced. If these factors are not considered the safety of the flight is put at risk, as the pilot's ability to manoeuvre and control the aircraft may be impaired.

All aircraft have a maximum allowable weight limit – as documented by the manufacturer – that cannot be exceeded, and specialist airline staff at Heathrow spend much time planning and calculating the expected load of a departing aircraft. They factor in every aspect, including passengers, baggage, cargo, crew, fuel, oil and onboard supplies such as catering.

Planners consider three key weights when preparing for a departure:

- The maximum take-off weight (MTOW), which has to be adjusted for local conditions at Heathrow based on the length of the runways, prevailing weather conditions, temperature, barometric pressure, wind direction and speed.
- The maximum landing weight (MLW) is

considered to ensure that the stress limits for the aircraft's landing gear are not exceeded. Attention is also given to the length of the landing runway (adjusted for local weather conditions), which may limit the distance the aircraft has to slow down and stop.
- The maximum zero fuel weight (MZFW) is the maximum structural loading on the wing root that may not be exceeded, and ensures the aircraft is not too heavy for the airframe. It is the total weight of the aircraft as prepared for flight less the weight of its fuel.

The final load information is then compared against these weights before the load sheet is approved and handed to the ground handling team, who must ensure that the aircraft is loaded exactly in accordance with the load plan, thereby ensuring the aircraft is suitably balanced for flight.

Push-back and start-up

Aircraft cannot reverse, so 'push-back' is a procedure during which an aircraft is literally pushed backwards away from one of Heathrow's gates by external power rather than manoeuvring under its own power. Push-backs are carried out by specialised low-profile vehicles called tractors or tugs.

LEFT A member of the ground crew releases the tow bar from an aircraft on stand. *(HAL)*

may be temporarily installed into the nose gear to disconnect it from the aircraft's normal steering mechanism. The pin prevents the aircraft from being mishandled by the tug – when overstressed the shear pin will snap, disconnecting the bar from the nose gear to prevent damage to the aircraft and tug. For obvious reasons it carries a 'Remove Before Flight' warning.

Once the push-back is completed, the tow bar is disconnected and the bypass pin is removed. The ground handler will show the bypass pin to the pilots to make it absolutely clear that it has been removed. The push-back is then complete and the aircraft can start taxiing forward under its own power.

Push-back tractors use a low-profile design to fit under the nose of the aircraft and need to be capable of generating significant levels of traction to move large passenger aircraft. The driver's cabin can be raised for increased visibility when reversing, and lowered to fit under the aircraft.

Conventional tugs use a tow bar to connect the tug to the nose landing gear of the aircraft. The tow bar is fixed laterally at the nose landing gear and moves vertically for height adjustment.

BELOW A Virgin Atlantic Airbus A340-600 being moved across the airfield is one of some 400 daily towing movements designed to maximise the efficiency of the available stands and space at Heathrow. *(Author)*

The procedure is subject to ground-control clearance confirming that it is clear to commence the procedure. Once clearance is obtained, the pilot communicates with the push-back tractor driver (or a ground handler walking alongside the aircraft in some cases) to start the push-back. To communicate, a headset may be connected to the aircraft near the nose gear.

Since the pilots cannot see what is behind the aircraft, steering is done by the push-back tractor driver and not by the pilots. Depending on the aircraft type and airline procedure, a bypass pin

It acts as a large lever to rotate the nose landing gear. Each aircraft type has a unique tow fitting, so the tow bar also acts as an adapter between the standard-sized tow pin on the tug and the type-specific fitting on the aircraft's landing gear.

As their name indicates, 'towbarless' (TBL) tractors do not use a tow bar. They scoop up the nose wheel and lift it off the ground, allowing the tug to manoeuvre the aircraft. This allows better control and higher speeds without anyone in the cockpit. The main advantage of a TBL tug is simplicity. By eliminating the tow bar, operators are relieved from maintaining a range of different tow bars. Connecting the tug directly to the aircraft's landing gear – instead of through a tow bar – also ensures better control and responsiveness when manoeuvring. This is advantageous at Heathrow where space is limited and there is a requirement for some 400 towing movements every day.

Before an aircraft is pushed back the area around it must be safe for the aircraft, its passengers and people working nearby. Often visible to passengers from the terminal are the large red and white blast screens (sometimes called 'jet blast deflectors') that are used to protect the terminals, the aircraft apron, roadways and walkways from the wash created by the exhausts of jet engines. They are designed to deflect the exhaust gases upwards. Jet blast deflectors began to appear at airports in the 1950s and grew in size as jet airliners such as the McDonnell Douglas DC-10 and MD-11, with engines mounted in the tail above the fuselage, became more commonplace.

Following push-back from the stand, aircraft pull forward to a minimum of 100m from the blast screen before the tug is disconnected. Engine starts on stands are limited to idle power on one engine, and special permission has to be obtained if more than one engine needs to be started, due to jet blast and the need to ensure the safety of ramp personnel.

Engine runs

Heathrow has three ground run enclosures (GREs), all owned by British Airways, which are used for engine testing of their own and other aircraft. The two latest GREs are designed for post-maintenance engine testing, including take-off power. After a jet engine has been overhauled or has parts replaced it is standard procedure to run the engine up to full thrust to test it. The run enclosures provide a safe and controlled environment for testing, and also one that minimises engine noise for nearby communities.

Aircraft that require engine runs above idle power that cannot be carried out in the run enclosures are sometimes given permission by the DMA to use a remote section of taxiway, although noise considerations have to be taken into account. An ASD team always attends these requests to position the aircraft safely into wind, to minimise engine stress and take into account blast effects and other hazards.

BELOW Three British Airways Boeing 747 aircraft flank the airline's new ground-run pen located in the aircraft maintenance area on the eastern side of the airport. *(HAL)*

ABOVE A British
Airways Boeing 777 is
de-iced at Terminal 5B
prior to departure. The
aircraft must depart
within a set time or it
will have to be de-iced
all over again. (HAL)

Aircraft de-icing

In freezing weather conditions the de-icing of aircraft is a crucial requirement, as a build-up of ice can jam an aircraft's control surfaces, preventing them from moving properly. Ice can also cause critical control surfaces to be rough and uneven, disrupting smooth airflow and significantly limiting the ability of the wing to generate lift, potentially causing an aircraft to crash. In addition, should large pieces of ice separate when the aircraft is in flight they can be ingested in the engines, potentially causing a catastrophic failure.

It's worth noting that de-icing differs from 'anti-icing', which refers to the process of protecting against the formation of a frozen contaminant in advance of adverse weather conditions. Anti-icing is accomplished by applying a protective layer of viscous liquid, called 'anti-ice fluid', over a surface to absorb the contaminate.

The point at which an aircraft needs to be de-iced is determined by the specifications set by the airframe manufacturer, and consequently varies depending on the aircraft type. The dew point is also taken into consideration.

De-icing involves the application of chemicals that not only de-ice, but also remain on the surface to delay ice from re-forming for a certain period of time. De-icing techniques are also used to ensure that engine inlets and various sensors on the outside of the aircraft are clear of ice or snow.

De-icing services at Heathrow are undertaken by the airline itself or an appointed third-party supplier, and are typically performed by spraying aircraft with a de-icing fluid based on propylene glycol, similar to the ethylene glycol anti-freeze used in some car engine coolants. De-icing fluids are always applied heated and diluted.

The fluids fall into two basic categories: heated glycol diluted with water for de-icing and snow/ frost removal, and anti-icing fluid which is an unheated and undiluted propylene glycol-based fluid applied to retard the future development of ice or to prevent falling snow or sleet from accumulating. In some cases both types of fluid are applied, first the heated glycol/water mixture to remove contaminants, followed by the unheated thickened fluid to keep ice from re-forming before the aircraft takes off.

The de-icing fluid is sprayed through a nozzle

that can be adjusted to give a moderate jet of fluid on to the contaminated area, and then a short time after the pressure is increased to use the hydraulic force to flush off loosened deposits.

Once an aircraft has been de-iced it is imperative that it leaves the gate within a set time limit, known as the 'hold-over time', otherwise there is a risk that the freezing conditions will create more ice and nullify the impact of the de-icing. Some thicker variants of de-icing fluid allow for longer hold-over times. The hold-over time is determined by a number of factors including air temperature, the type of aircraft in question, the de-icing product used and the water content of that fluid. Aircraft that haven't left within the required window and have to be de-iced again add to congestion and impact on airport efficiency.

HEATHROW PEOPLE: SAJID AWAN

Dispatcher for Emirates Airlines

I've been working at Heathrow since 1994, when I started as a check-in agent with Scandinavian Air Services. Over time I have worked my way up and I'm now a Dispatcher working with some of the most sophisticated aircraft and systems, including the Emirates A380.

Getting the world's largest commercial aircraft dispatched on time is no mean feat. All of Emirates' flight planning is done from the airline's headquarters in Dubai, so my job is largely focused on acting as the link between what is happening on the ground at Heathrow and at HQ in Dubai. There is a regular flow of information between the two offices to make sure the flight is being prepared according to the airline's policy.

Some 24 hours in advance I receive the passenger list and from that I can make some reasonably clear assumptions on the likely baggage load. A full A380 will normally have 600–700 bags, packed into 22 containers. Three hours before the flight I receive the cargo manifest from Emirates Cargo and I also take receipt of the flight plan prepared in Dubai, specifying the amount of fuel needed for the flight.

The final part of the jigsaw puzzle comes in the form of the load plan from HQ. It's essential for me to get this to the ramp team from ground-handling company Dnata as soon as it's available, as it explains exactly how the aircraft needs to be loaded so that it is balanced correctly prior to take-off. The load sheet also contains special instructions in relation to any 'special cargo', such as live animals.

I pop my head into the cockpit to make sure the captain is briefed on any dangerous cargo items being carried (these can include the likes of flammable paint and dry ice used to keep perishable fruit and vegetables cool in the hold).

There is a real buzz on the ground as final preparations get under way and 507 passengers start boarding. Fortunately everyone turns up on time as late-running passengers can seriously delay the departure as their bags would have to be unloaded. With the first-class and transfer passenger bags located close to the cargo door for maximum efficiency when landing in Dubai, the bags we could be looking for would be deep in the hold.

With the passengers aboard and the bags and cargo loaded the flight is closed and the cumulative weights are used to calculate the 'actual zero fuel weight'. This figure is passed to the captain, who can adjust the final fuel load if he feels it's necessary. With the doors closed the air bridge can be removed and the flight is ready for departure.

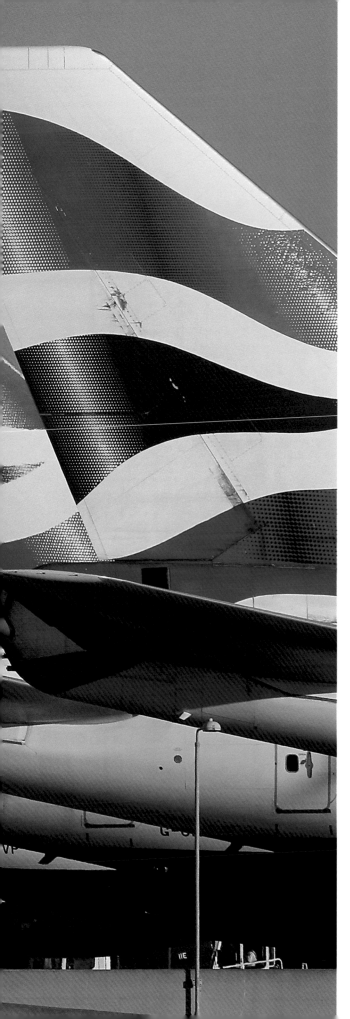

CHAPTER 9

British Airways

OPPOSITE British Airways aircraft on stand at Terminal 5.
(Author)

ABOVE The BA Headquarters building, Waterside – planning for every one of the airline's flights starts here up to two years before the actual departure date. *(Author)*

Introduction

British Airways (BA) is one of the world's leading airlines and is synonymous with Heathrow. The airline uses Heathrow as its principal UK base, with its distinctive red, white and blue tailfin design borne by approximately 50% of the flights to and from the airport.

British Airways is part of the International Airlines Group (IAG), and was a founding member and an integral part of the successful 'oneworld' alliance.

BA employs some 40,000 people, based mainly in the UK, although it has staff based in over 75 countries around the world. The airline has 15,000 cabin crew and 3,500 pilots, and its fleet consists of some 270 aircraft, about half of which are used for long-haul flights. It operates in excess of 300,000 flights per annum, with a route network to more than 160 different cities.

BA is the flag-carrier airline of the United Kingdom, and the largest airline in the country based on fleet size, international flights and international destinations.

Waterside

BA's international head office, Waterside, is located just outside the airport perimeter on Harmondsworth Moor. As Peter Lynam, BA's Head of Network Operations, explains, 'Every BA flight within our network starts out life at Waterside, and our Operations Centre forms the overall hub for the complete British Airways global operation. This is where every flight – not just those operating from Heathrow – is planned, promoted and approved.'

Construction of Waterside started in late 1995

and it was completed some two-and-a-half years later. The impressive building houses a health centre, hairdressing and beauty salon, travel centre, supermarket, bank, restaurants and cafes and a 400-seater auditorium.

Today around 3,800 people work at the BA headquarters in a range of roles including sales, marketing and promotion of flight offers, as well as a revenue management team which sets ticket prices, looks at the best use of capacity and works out how to optimise fares. There is a significant procurement department purchasing everything from actual aircraft to sausages for in-flight catering.

There is also a significant finance department dealing with the affairs of what is a £10 billion company. Much time is spent dealing with foreign exchange issues, since the airline is buying fuel in multiple foreign destinations on a daily basis and having to pay for it in local currency. Other departments include security, legal, human resources, recruitment, IT and government relations staff.

Waterside houses a group of engineering staff who work closely with the core team based at the BA Engineering facilities on the east side of the airfield. At Waterside, engineers study live telemetry from aircraft with a view to pre-empting any technical issues.

While Heathrow's management team have to deal with incidents at or close to their own airport, airlines such as BA can potentially have an incident happen anywhere in the world, given that their fleet of aircraft are constantly on the move. In fact, a BA aircraft takes off every 75 seconds somewhere in the world.

Fortunately accidents are very rare occurrences, but an 'away from home' incident is all the more challenging to deal with and requires the airline to have highly trained staff and a sophisticated crisis control centre with state-of-the-art communications technology. Waterside houses BA's Crisis Management Centre (CMC) – this is the corporate command and control centre in the event of a major incident occurring either at Heathrow or elsewhere in the world. It also provides policy guidance, responds to enquiries from the media and has access to a wide variety of resources including key personnel and communication aids. The CMC seats 33 key members of BA's crisis management team, providing them with real-time information. Audio-visual systems in the room are streamlined to manage and display

35 data and 15 video sources linking in all of BA's operational sites globally.

A separate Emergency Procedures Information Centre (EPIC) co-ordinates and controls all passenger and family information for the airline and liaises with telephone enquiry centres worldwide. It provides information and support to CMC.

Aircraft

Of the 270 operational aircraft currently in the BA fleet, around 220 are based at Heathrow, this number having increased in July 2013 with the addition of the first of a new fleet of 12 Airbus A380s. The Airbus fleet will initially fly from Heathrow to Los Angeles, Hong Kong and Johannesburg.

The A380 will be used to periodically phase out the older Boeing 747-400 fleet that has been the airline's long-haul workhorse since it entered service in 1989. The A380's innovative design makes it much quieter during take-off and landing and 16% more fuel-efficient than a Boeing 747. Of the 12 that British Airways ordered for delivery by 2016, three arrived in 2013 and a further five are due in 2014.

The airline has also taken delivery of the Boeing 787 Dreamliner. Willie Walsh, Chief

ABOVE One of 55 Boeing 747-400 aircraft in service with British Airways taxis towards the runway. *(Author)*

BRITISH AIRWAYS AIRCRAFT FLEET (AS AT JANUARY 2014)

Aircraft	In service	On order	Passengers*	Length (m)	Wingspan (m)	Cruising speed (mph)	Range with max payload (miles)
Airbus A318–100	2		32	31.4	34.1	511	3,600
Airbus A319–100	44		113–132	33.8	34.1	511	4,200
Airbus A320–200	46		135–162	37.5	34.1	511	3,800
Airbus A321–200	17		153–188	44.5	34.1	511	3,500
Airbus A350	0	18	To be announced	64.8	64.8	561	9,300
Airbus A380–800	3	9	469	72.7	79.5	587	9,755
Boeing 737–400	19		139–153		28.8	485	2,700
Boeing 747–400	55		291–345	76.3	64.4	575	7,300
Boeing 767–300ER	21		189–259	54.9	47.6	530	6,890
Boeing 777–200	3		216	63.7	60.9	590	5,240
Boeing 777–200ER	43		219–283	63.7	60.9	590	7,725
Boeing 777–300ER	6	6	297	73.9	64.8	590	7,930
Boeing 787–8	4	4	214	56.7	60.0	567	9,100
Boeing 787–9	0	16	To be announced	62.8	60	567	9,100

* Depending on the seating configuration of the aircraft.

Executive of IAG, says: 'The 787 is a tremendous, innovative aircraft which sets new standards for environmental performance and operating efficiency.' The airline has announced plans to convert further options for the aircraft into firm orders for delivery from 2017, including for the new, larger 787-10 version.

Upon arrival, both aircraft types began a complex 'entry into service' programme, which saw pilot and cabin crew training taking place at Manston Airport in Kent, customer service trials at Heathrow, and short-haul flying for both aircraft.

Further modernisation of BA's fleet between 2017 and 2023 will see 30 of its 747s replaced by 24 787 Dreamliners and 6 additional Boeing 777-300ERs.

Orders for 18 further Dreamliners and the same number of Airbus A350s (which completed its maiden test flight in June 2013) are also in place.

Network operations

Peter Lynam, BA's Head of Network Operations, scans the screen in front of him and in seconds can tell that the first long-haul outbound flight of the day, BA117 to New York, a Boeing 747-400, is about halfway across the Atlantic en route to JFK International Airport. He can also see the flight has 227 passengers, 8 children, 2 infants, 2 pilots and 14 cabin crew aboard. The aircraft is carrying 7,600kg of cargo and 3,700kg of baggage. The

on-screen information also tells him exactly where the cargo is placed in the hold and how the aircraft was balanced. Also on board at departure were 82,868kg of fuel, including four tonnes added by the captain as a reserve. Looking at the exact time the last door closed, Lynam is frustrated the flight departed Heathrow 14 minutes late. With a flight time of 6 hours and 58 minutes and fortunately only an eight-knot headwind, BA117 is predicted to land just 11 minutes late.

At the heart of BA's headquarters, and perhaps best described as the 'custodian' of the airline, is the impressive Network Operations Centre (NOC), which plays the key role in BA's operational performance. The NOC is undoubtedly the airline's nerve centre, and the scale of the challenge begins to emerge considering that somewhere in the world a BA plane takes off or lands every 75 seconds. As if that isn't complicated enough, there are numerous other factors to consider, from aircraft maintenance requirements to flight-crew preferences. The teams within the NOC find the solutions to resolve these issues and more on a daily basis.

'On average, we have around 340 inbound and 340 outbound flights at Heathrow alone every day. That makes for 680 complex operations that have to work well,' says Lynam, who has worked in various roles at BA for the past 36 years.

'We like to start the day with a "clean slate",' he adds, 'but we know there are a variety of things that can happen which impact on our schedules

and cause disruption.' Lynam's extensive team of planners and controllers are trained not only to operate the airline in the most feasible and efficient manner and maintain the integrity of the airline's published schedules, but also to deal with a host of contingencies. These can range from a simple delay to one aircraft in the fleet to the knock-on effects of an air traffic controllers' strike in France that requires aircraft to be diverted around French airspace. This can cause delays, meaning aircraft must use more fuel, and it places an unusually high demand on the surrounding airspace, which is loaded to cope with the extra traffic. The team also has to deal with disruption caused by severe weather, security scares or aircraft taken out of service at short notice for maintenance reasons. 'If you start the day with an issue, it's inevitably only going to get tougher, so our systems are designed to "expect the unexpected" and offer as much predictability as possible,' he says.

To try and ensure as smooth and effective an execution of BA's 680 daily Heathrow movements as possible, flights are scheduled and planned 18 months ahead of time. 'This business involves

ABOVE British Airways' first Boeing 787 Dreamliner arrives at Heathrow in June 2013. *(Jeff Garrish/BA)*

LEFT A British Airways Boeing 777 takes off from runway 27R with Terminal 5C visible in the background. The airline operates more than 50 777s in different variants. *(HAL)*

a very significant amount of forward planning and strategy,' says Lynam, 'and we work up to a year-and-a-half in advance by looking at customer demand, the airline's market share, the product [seating configuration] we have on offer, as well as the political situation in the countries we fly to and how that might change.'

Major events like the Olympics, or significant political upheaval such as the 'Arab Spring' in late 2010, can have a major impact on flight planning. Added into the mix is the complex nature of long lead-times, with due consideration having to be given to procurement of fuel, spares, crew, stands, slots, pilots, cabin crew, catering and, of course, aircraft. The planning team works closely with BA Engineering, since the aircraft need to undergo regular maintenance, which can vary from simple daily checks to complex ones that may take an aircraft out of operation for several weeks.

'The whole planning process becomes more and more refined the closer we get to each individual flight,' adds Lynam, 'and throughout the process there is emphasis on ensuring that every part of the planning for each individual flight is fully feasible. It's a constant balancing act between demand from our customers – who can book almost a year in advance – and the supply of flights provided by the airline.'

To keep on top of what is certainly one of the most complex logistical challenges in the world, BA rely on a number of sophisticated IT systems, including databases that maintain detailed records of the airline's 'assets', including flight and cabin crew, aircraft, spares and customer records. Running in tandem with these are so-called 'Optimiser' systems that are used to shrink

resources and make the airline as efficient as possible based on the information provided by the databases. Systems such as these are hugely complex and are typically renewed on a ten-year cycle, costing the airline anywhere from £5 to 10 million to implement.

'In years gone by our IT systems were all about capturing data, but that has become so much simpler and our systems are now used to provide solutions to balance the airline's business to get the best results. Think of the combined systems as a "graphic equaliser" that can be tuned to achieve the best possible results,' says Lynam. From ba.com to the use of iPads on board, to systems that optimise the use of fuel, to exploitation of the latest mobile technology, IT systems play a vital role in the airline's daily operations.

Optimisation is also done with the aircraft fleet, with lead-times built in for the turnround procedures, given the known congestion issues at Heathrow. Delayed turnrounds have significant cost implications, with the airline having to provide for spare aircraft, flight and cabin crew as well as spare stand capacity to be used in the case of disruption. Around ten aircraft are typically scheduled for maintenance and a further ten held in reserve for contingency. If these fallback positions fail to resolve a problem, the airline can look at consolidating two flights into one (for example, turning two A320 departures into a single, larger flight using a Boeing 777) to minimise disruption for passengers, or may consider leasing a spare aircraft from a third-party supplier.

Four key departments make up the NOC and are supported by BA's own ATC Operations and Customer Service Control departments:

- Centralised Load Control.
- Flight Technical Dispatch.
- Tactical Planning.
- Aircraft and Crew Planning.

Centralised Load Control

Determining the balance and weight of an aircraft filled with passengers, baggage, cargo and fuel is a fundamental part of any flight, and is done to ensure the aircraft is not too heavy for the engines to lift it off the ground. Linked to this is the fact that the undercarriage can only support a certain amount of weight on landing. Furthermore, the balance of the aircraft is crucial, as the location of the passengers, baggage, cargo and fuel will

BELOW Two British Airways de-icing crews tackle a Boeing 777 during the heavy snowstorms that blighted the airport in 2010. *(HAL)*

affect the aircraft's trim. Loading too much weight into a particular part of the aircraft could cause damage and shorten its working life.

So, well before a flight departs the Centralised Load Control team take all of the available weight data – including notional passenger weights, baggage estimates based on online check-in information, cargo, catering, fuel and other onboard supplies – to build a 'base load', which is then updated as remaining passengers check in at the airport or last-minute cargo is added.

Given its cost, careful consideration is given to the fuel load. At around £650 per tonne, and with a Boeing 747 heading to Singapore likely to use around 160,000kg, that's about £104,000 for the one flight alone, making fuel expenditure one of the airline's single biggest expenses. On average BA incurs costs of some £10 million per day on fuel, or some £3.5 billion a year.

With complete information to hand, the team generates a load sheet and balance chart, which is then sent directly to the flight deck, giving the pilots the accurate information they need prior to departure. The pilots sign off and approve the information before taking off.

Flight Technical Dispatch

The Flight Dispatch team takes several factors into account, including the likely flying time, the load to be carried, the weather en route as well as the most cost-effective airspace to fly across – Scottish airspace, for example, is more expensive than Irish. For a transatlantic flight consideration is given to the jetstream and also which of the ten main routes are preferred for the flight in question. The team makes available all navigation, performance and other necessary operational data to the flight crew, to ensure safe air and ground operations. Once all of this information has been assessed, a flight plan is then filed with the Centralised Flow Management Unit (CFMU) in Brussels, who then co-ordinate its allowable start time (slot) with the various national and local ATC units across Europe and its close neighbours.

Tactical Planning

The Tactical Planning team provides the vital link between the strategic planning conducted 18 months earlier and the minute-by-minute decision-makers dealing with flights on a day-to-day basis. Their planning horizon is based on flight operations in the coming ten days. This team refines its planning every 24 hours and constructs a schedule to determine where conflicts might occur, to ensure that there is sufficient turnround time and that the various resources needed for each flight are going to be available to meet requirements.

Aircraft and Crew Planning

The final cog in the wheel is the Crew Planning team, which maintains and controls feasible crew plans to meet the demands of the airline. They do so by making optimum use of resources against a background of worldwide commercial, operational and legal constraints and complex and variable crew agreements.

At any one time around 5,000 of the 15,000-strong BA cabin crew will be on duty, and the planning team need to ensure the airline doesn't fall foul of working-time regulations and CAA guidelines, which set out strict parameters for crew and pilots.

A sophisticated computer system produces 170 million roster options, giving the airline significant choices depending on current objectives, be those based on cost-effectiveness, optimising crew lifestyles or improving punctuality. Planners convert the proposed flying programme and construct what is called a 'crew pairing', with both flight and cabin crew then assigned to an agreed roster for the month ahead. The crew must be both legal and qualified to operate these pairings. It is not uncommon for the rosters to undergo many changes due to the dynamic nature of the airline business, with additional flights being planned, upgrades of aircraft on certain routes to satisfy the supply and demand curves of the business, or aircraft becoming unserviceable resulting in aircraft changes to ensure the published schedule can be flown.

Once a flight from Heathrow has taken off, the NOC can be in direct contact with the aircraft using satellite phones or phone patches via commercial HF radio links and sharing information using the Aircraft Communications Addressing and Reporting System (ACARS), a digital datalink system for transmission of short, relatively simple messages between aircraft and ground stations via radio or satellite.

'Once airborne we (and most airlines in Europe) tend to leave the decision-making in the hands of the pilot, but do obviously keep in regular contact with the aircraft,' says Peter Lynam. 'This is a different system to the one used in the USA. Our US colleagues use a so-called "licensed dispatcher" who has shared responsibility for the flight. He tracks it for its duration until it lands, participating in decision-making during the flight, effectively giving the aircraft one captain in the air and one on the ground.'

In the event of an incident – which can be relatively minor in nature or catastrophic, resulting in loss of life – the airline has a duty to provide care to its passengers, crew and relatives of those involved, and again the NOC teams are at the heart of this. In instances similar to the BA762 flight to Oslo described on page 118, the NOC receives a live feed of information from the flight crew. Physically dealing with the emergency situation is left in the capable hands of the Airside Safety Department and the airport's emergency services, while at the NOC resources are allocated to manage the incident and to keep the airline running. The information to hand is compared to previous similar situations to enable the airline to get a sense of the likely knock-on effect of the incident.

'In this particular instance [BA762], we came to the conclusion that the airport would likely be shut for a potentially protracted period of time while the incident was dealt with,' says Lynam. 'Obviously our immediate concern was about the flight in question, but we also had to consider the knock-on effect through the day for other flights. It becomes vitally important that we as an airline make a quick and clear decision on how to deal with the remainder of the day, and then communicate with our customers to prevent them coming to the airport unnecessarily.' Overcrowding in any one of Heathrow's terminals will cause a potential health and safety issue, so the airline needs to take a quick and clear decision in the best interests of all concerned. In a suspected security situation, passenger baggage is generally impounded for further investigation and the airline then faces the challenge of reuniting passengers with their bags.

Cabin crew

When people think of BA, the faces they see are usually those of the airline's cabin crew. From a warm welcome onboard to a sincere thank you and goodbye at the end of the flight, the cabin crew play a vital role in defining a customer's experience of British Airways.

The airline currently has four cabin crew fleets, three of which are based at Heathrow. These are the Worldwide, Euro-Fleet and Mixed Fleet crews,

with a fourth contingent based at Gatwick Airport. The crews travel the world on one of the most extensive route networks of any airline.

BA also has international cabin crew based in Buenos Aires, Mexico City, São Paulo, Cairo, Mumbai, Tokyo, Hong Kong, Delhi, Bahrain, Singapore, Chennai and Bangalore. The aim here is to provide customers travelling these routes with local crew who have first-hand knowledge of their language and culture.

Once a month each one of BA's 3,500 pilots and 15,000 cabin crew receive a schedule listing the routes they are rostered to fly over the coming 30 days. Crew are allocated to specific aircraft types.

British Airways Engineering

Heathrow plays host to the operational home of British Airways Engineering (BA Engineering), responsible for managing some 2,000 aircraft turnrounds each week courtesy of a comprehensive range of facilities, equipment and some of the most experienced aviation engineering personnel in the world.

Located on the eastern perimeter of the airport, it's a fascinating environment that sees some of the latest aviation technology being worked on in some of the airport's most historically significant buildings. BA Engineering employs some 5,500 personnel globally, with more than half of that team based at Heathrow across different shifts and work areas.

BA Engineering provides a range of services including periodic and scheduled maintenance, the refurbishment, conversion and enhancement of cabin interiors, aircraft storage, aircraft recovery and support, landing gear and engine replacement, component overhaul, regulatory inspections, aircraft painting and the application of decals and graphics, as well as major and minor structural modifications and repairs.

'We offer what's called "nose to tail capability",' says BA's Director of Engineering, Andy Kerswill, 'which allows us to carry out aircraft maintenance and component repair and overhaul using state-of-the-art equipment, avionics and mechanical workshop facilities.

'It's imperative when we have an "Aircraft on Ground" (AOG) situation – which indicates that an aircraft has a problem serious enough to prevent it from flying – that we need to act quickly. This can happen either at base [Heathrow] or with

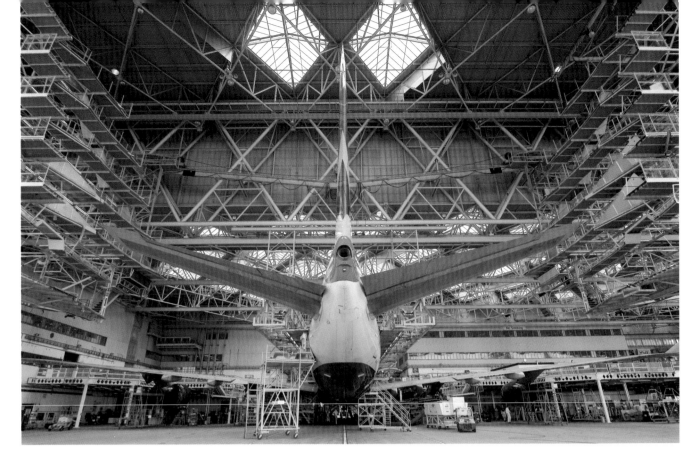

an aircraft at one of the airline's destinations, so it's imperative that we put the aircraft back into service and prevent further delays or cancellations of the planned itinerary.' BA Engineering carries a significant inventory of Boeing and Airbus spare parts that can be called on by BA or by customers using the same aircraft types.

One of the most essential and fascinating activities that takes place at BA Engineering happens in the Engine Health Monitoring (EHM) department, where each and every BA engine goes through a rigorous check. Much of the team's work is based on information gathered from magnetic chip detectors (MCDs). These small plugs are installed by the engine manufacturer in the oil systems of their engines and have the potential of saving the airline hundreds of thousands of pounds by detecting problems early on.

Like any engine, over a period of time wear and tear causes microscopic metal chips to break loose from engine parts and circulate in the engine oil. The MCDs pick up these chips, thereby acting as an early warning system, following which they are sent for analysis in the EHM laboratory. MCDs are also checked as a matter of course during routine maintenance.

EHM Team Leader Trevor Ford says: 'This is an integral and essential function for the BA Engineering operation and vital to the aircraft's safety and to prolonging the life of the engines. When we receive the MCDs we remove any

excess oil with degreasing solvent and make an initial inspection under a microscope to see what we're dealing with. We take a sample from every MCD whether there is an issue or not, as this allows a thorough history of each engine to be maintained.'

MCDs are present on every engine in the fleet, but the actual number depends on the engine type. Modern engines have more than older engines – each Rolls-Royce Trent 1000

on the Boeing 787 engine has 10 MCDs (20 per aircraft), while each Rolls-Royce Trent 900 on the Airbus A380 has 9 per engine (36 per aircraft). In addition, each of these engines – as well as many engines on the 777 fleet – also contains a sophisticated electronic sensor which acts as the 'master MCD', the purpose of which is to indicate the possibility of debris ahead of undertaking a visual inspection. Each MCD is colour-coded to a specific position on a particular engine, which makes the team's job easier both when analysing and subsequently replacing the detectors.

When chips have been detected a detailed analysis is done using a spectro-analyser that interprets the elemental make-up of the particles. 'The analysis establishes the make-up of the chips, which are then compared using complex visualisation software in the form of a graphic overlay against the standards provided by the engine manufacturer,' says Ford. 'This allows us to quickly identify from which component the chips could have come. From here we can make an informed decision as to whether what we have found is within acceptable manufacturer guidelines and, if not, whether the aircraft needs to be taken out of service either at a point in the near future or immediately, if it is a serious issue.

'Obviously it's preferable not to take the aircraft out of service unnecessarily and, if possible, to deal with the issue when the aircraft is with us for routine maintenance, but we won't hesitate to ground an aircraft if it is warranted.' Every aircraft is signed off and approved before being put back into service.

BA Engineering also has an extensive machining capability and a full paint shop facility. 'We repaint every BA aircraft on a five-year rotation to ensure the identity of the BA brand is both consistent and of the highest standard the world over,' says External Appearance Manager, Dave Barnes. 'It's a fairly complex exercise, painting something the size of a Boeing 747, but with the use of some clever templates and lots of past experience we've got it down to a fine art,' he says with a wry smile.

The engineering team also deals with avionics and maintains a broad range of electrical, electronic and mechanical components, as well as diagnostics technology for fault-finding.

BELOW Tail fin design template for a British Airways Boeing 767-300. (Author)

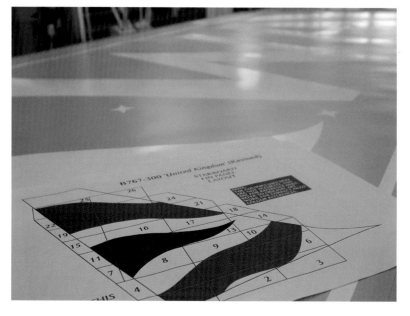

ABOVE, RIGHT AND BELOW British Airways Engineering carries out a wide range of maintenance services on everything from engines, to control surfaces and passenger seats. *(Author)*

BELOW AND RIGHT A spectrum analyser is used to interpret the elemental make-up of minute particles found in engine oil. *(Author)*

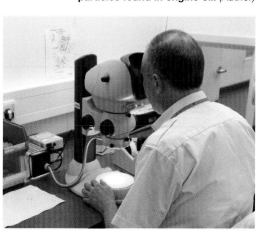

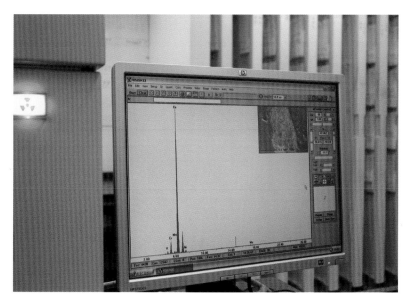

ABOVE Two of British Airways' flight training simulators. *(Jonathan Harvey/BA Engineering)*

RIGHT Flight crew train on the A380 simulator. *(Nick Morrish/BA)*

PREPARING FOR THE ARRIVAL OF THE A380

In order to prepare for the arrival of their fleet of A380s, British Airways Engineering had to complete a series of trials in their facility at Heathrow – a landmark 1950s listed building – which had to be specifically converted for the world's largest passenger airliner. The trials involved putting the hangar through its paces, and Airbus flew in one of its 'super jumbos' to facilitate the test.

During the refurbishment of the four-hangar complex, , the team had to overcome several challenges, one of which was to slot the giant aircraft's 24m-high tail fin into the hangar. More than 138 tonnes of steelwork was added to the roof structure to create a slot for the tail to pass through.

The upgraded state-of-the-art facility means British Airways is able to maintain its own A380 fleet as well as those of other carriers. There are many other considerations in addition to the infrastructure requirements. For example, a team of flight operations engineers has been spending time working on flight progress charts. These maps show the air routes and diversion airports and determine which airfield a flight crew can divert to in the event of an emergency. They are ranked from 'primary preferred' down to 'emergency only'. 'Development of flight progress charts for the A380 is an interesting challenge as there are fewer airports that can accept it,' says engineer Anil Padhra. 'For some it would present an operational disruption if it were to land.'

BELOW British Airways cabin crew welcome the arrival of the airline's first A380 at Heathrow in July 2013. *(Author)*

BAE's engineers need to be at the cutting edge, with ever-increasing levels of proficiency when new technology comes to the fore, such as Boeing's Dreamliner 787. The team's scope of work extends to the flight desk, galley, oxygen supply, radio and radar, flight management control functions, lighting and wiring looms. Other teams focus on specific areas, including mechanical flight controls, wheels, brakes, radomes, control surfaces, engine cowls and thrust reverser repairs.

Engineering's responsibility also extends to cabin interior and safety services, including seating on the flight deck and for passengers, carpeting, evacuation slides, life vests and survival packs used in the event of an emergency.

The arrival of the Airbus A380 and Boeing 787 have been major milestones not only for the airline but also for BAE. 'These are exciting times as we welcome two state-of-the-art aircraft types into our fleet and balance the needs of these against an older fleet that requires more maintenance effort,' says Kerswill, whose engineering team have spent months preparing for the latest additions to visit BAE's hangars.

Flight training

B A provides a range of training for flight and cabin crew at Heathrow. At around £10–12 million each, BA has made a significant investment in 16 flight simulators for its flight training operation. The simulators cover both Boeing and Airbus aircraft types and have recently been consolidated from the airline's nearby Cranebank Training facility into a new flight training campus and centre of excellence at its maintenance base.

This dedicated campus is set to become one of the world's leading flight training facilities for the global aviation industry. BA also has a Ground School facility that provides integrated computer-based training with fixed base flight simulators. In support of the overall operation, a team of dedicated simulator engineers provide 24-hour coverage and maintenance to the highest standards, based on previous experience working within a full flight simulation environment.

BA also provides some of the most comprehensive certified cabin crew training in the world, enabling staff on board aircraft to deal with a complete range of emergency situations including fire and smoke, decompression, emergency landing and ditching.

HEATHROW PEOPLE: NEIL AND CLAIRE PARSONS

BA cabin crew

Unlike most people who start their working day in the morning, today we are driving to work when the tired workforce are heading home. The time in the car is 17:15 and, unusually, we're early! We are heading south, down the M40 from our home in the Cotswolds to Heathrow.

I say 'we', because my husband and I work together. I joined British Airways' World Wide Cabin Crew in October 1997, which is where I met my husband Neil. We are one of many couples who are fortunate to fly together on a 'married roster', which the airline recognises and supports. We do have a young daughter but thanks to our devoted and dedicated parents we manage to balance days like today and normal family life with considerable ease. I work as a member of the main crew and Neil is a Customer Service Leader, in charge of his respective cabin.

This evening we are reporting for a nine-day Singapore and Sydney trip, being operated on a Boeing 777-300 aircraft. Ninety minutes before the scheduled departure we attend a flight briefing, run by the Cabin Service Director. We spend 20 minutes reviewing our specific aircraft's safety procedures, equipment and medical provisions before finally choosing our working positions, which is done on a seniority basis.

Our flight time today to Singapore is 12 hours and 50 minutes, making this one of the longest routes we operate. For me, the flights seem to go so quickly when we are working; we do also get to have a break, including a lie down in the bunk area at the rear of the aircraft. This is a legal requirement on long-range sectors.

We're aware that many people see our job as 'tea or coffee?', 'chicken or beef?' and yes, we do serve these on board, but our training is predominantly about safety, equipment, procedures, life rafts, quick cuffs (used in the unlikely event of having to restrain an unruly passenger), dealing with bomb threats and medical training.

Once on board we immediately check our safety equipment and the cabin to make sure that not only is everything in its place for our customers (blankets, headsets and magazines), but also for any items that are suspicious, which might be deemed a security threat.

During the flight we have to placate two customers who had a dispute over their seating, and a team of us have to perform resuscitation on a female passenger who collapsed and went into full cardiac arrest. Granted, this was one of our more challenging flights, but our customers disembark with positive comments, which is always gratifying. We are proud of what we do and the adventures that it takes us on together.

(Nick Morrish/BA)

Cargo operations

OPPOSITE A Singapore Airlines Airbus A380 is loaded with belly-hold cargo at Terminal 3. The airliner has 184m³ of bulk and container freight volume. (HAL)

Introduction

Heathrow is the largest UK port in terms of exports by value to non-European Union countries, and is regarded as the country's most important airport in terms of cargo tonnage, handling some 64% of the UK total. It's also interesting to note that over a quarter of all exports from the UK to Brazil, Russia, India and China go through Heathrow. On average each flight to these countries is worth between £400,000 and £1 million. This makes freight going through Heathrow alone worth an estimated £35 billion a year.

Aviation industry forecasts estimate that air freight traffic volumes are likely to more than triple over the coming 20 years, with 80% of air cargo traffic likely to be long-range, intercontinental, hub-to-hub services, and the airport is well placed to take advantage of its leading position in the cargo sector.

Of all the freight departing from Heathrow, around 95% is carried 'belly-hold' (*ie* on flights that also carry passengers), with the remaining 5% carried by dedicated freighters. The airport has 22 on-airport warehouses, the largest of which, by a considerable margin, is IAG Cargo (part of the wider International Airlines Group), which runs a remarkable logistics operation at Heathrow.

The Cargo Centre was originally opened by British Airways in 1996 but has since seen the consolidation of the cargo operations of British Airways, Iberia and BMI into a single business. Today IAG Cargo is the world's seventh-largest international cargo airline and is a core component of the British Airways business. A worldwide operation with additional UK stations at Manchester, Birmingham and Glasgow Airports, the airline carries almost a million tonnes of freight, mail and courier shipments every year.

The cargo terminal itself – one of the most impressive facilities of its kind anywhere in the world – is located south of the two runways, towards the south-western corner of the airfield, and is capable of handling more than 800,000 tonnes of cargo a year. A cargo tunnel connects it to the central terminal area while the Western Tug Road provides a dedicated link to Terminal 5.

The future for IAG Cargo and its impressive facility is certainly a bright one, but its growth too is constrained by the airport's capacity issues. In its submission to the Airports Commission, the Freight Transport Association (FTA) argues that Heathrow should be permitted to expand its operations through additional runway capacity in order to meet existing and future demand from international trade. It argues that while Heathrow competes as a mixed-use hub with its Continental competitors in Paris, Frankfurt, Amsterdam and Madrid, a lack of available slots at the airport is constraining its ability to add new routes to emerging markets, critical to the UK's economic growth.

BELOW Aerial view of the World Cargo Centre located on the south side of the airport. *(HAL)*

Cargo volumes

A comparison of the world's busiest airports by cargo traffic shows that London Heathrow is the world's 16th busiest cargo airport by total cargo metric tonnes, and the third busiest European airport for cargo.

Rank	Airport	Location	Cargo volume (metric tonnes)
1	Hong Kong International Airport	Chek Lap Kok, Hong Kong, China	3,976,768
2	Memphis International Airport	Memphis, Tennessee, United States	3,916,410
3	Shanghai Pudong International Airport	Pudong, Shanghai, China	3,085,268
4	Ted Stevens Anchorage International Airport	Anchorage, Alaska, United States	2,543,155
5	Incheon International Airport	Incheon, Seoul National Capital Area, South Korea	2,539,221
6	Paris–Charles de Gaulle Airport	Seine-et-Marne/Seine-Saint-Denis/Val-d'Oise, Île-de-France, France	2,300,063
7	Frankfurt Airport	Flughafen (Frankfurt am Main), Frankfurt, Hesse, Germany	2,214,939
8	Dubai International Airport	Dubai, United Arab Emirates	2,194,264
9	Louisville International Airport	Louisville, Kentucky, United States	2,188,422
10	Narita International Airport	Narita, Chiba, Kantō, Honshō, Japan	1,945,351
11	Singapore Changi Airport	Changi, East Region, Singapore	1,898,850
12	Miami International Airport	Miami, Florida, United States	1,841,929
13	Los Angeles International Airport	Los Angeles, California, United States	1,696,115
14	Beijing Capital International Airport	Chaoyang, Beijing, China	1,640,247
15	Taiwan Taoyuan International Airport	Dayuan, Taoyuan, Taiwan, Republic of China	1,627,463
16	London Heathrow Airport	London, United Kingdom	1,569,449
17	Amsterdam Airport Schiphol	Haarlemmermeer, North Holland, Netherlands	1,549,686
18	John F. Kennedy International Airport	New York City, New York, United States	1,344,537
19	Suvarnabhumi Airport	Racha Thewa, Bang Phli, Samut Prakan, Greater Bangkok, Central Thailand	1,321,853
20	O'Hare International Airport	Chicago, Illinois, United States	1,311,622

Source: Airports Council International, 2013.

IAG Cargo Centre

More than 1,500 people work at the IAG Cargo Centre at Heathrow, which is owned and operated by IAG Cargo, the single business unit created in April 2011 following the merger of British Airways World Cargo and Iberia Cargo. Their network spans over 350 destinations, and in 2012 the business generated commercial revenue in excess of £1 billion.

IAG Cargo has two main facilities on the Heathrow campus:

■ Ascentis – the major freight-processing centre.
■ Premia – the premium handling centre focused on moving specific products swiftly and efficiently through the building.

Ascentis is one of the largest and most advanced automated freight-handling systems in the world, with around 70% of all Heathrow freight passing through this facility. It was designed to provide the most efficient operational environment for cargo handling, enabling the team to design new business processes and cargo systems to improve operational performance even further.

Other points of interest inside the facility include:

■ A fully operational mortuary designed to cater for the sensitive movement of human remains.
■ Special fridges for meat and fish products to be kept at the right temperature.
■ A special handling centre managing time-critical services for couriers and express delivery customers.
■ Two dedicated airside stands for freighters.

Premia officially opened in September 2006. The facility was originally 7,200m² (77,000ft²). However, following significant investment capacity was expanded by 20% in 2012.

In 2013 a new temperature-controlled facility was erected between the Ascentis and Premia buildings to cater for the burgeoning demand in pharmaceutical products.

Adjacent to the IAG Cargo facility is the Royal Mail's Heathrow Worldwide Distribution Centre (HWDC), which processes both export and import mail to and from the UK. It handles around eight million pieces of export mail and nine million pieces of import mail every week.

Aircraft and equipment

Given the strength of the network operated by IAG through British Airways, Iberia and BMI, the vast majority of goods are carried in the belly hold of passenger aircraft. Up until April 2014 IAG used to 'wet lease' three Boeing 747-8 freighters to serve key trade routes including Johannesburg, Hong Kong and Madrid. Since then IAG has entered into an agreement with Qatar Airways to provide Boeing 777F freighter capacity between Hong Kong and London.

A 'wet lease' is a leasing arrangement whereby one airline provides an aircraft, the complete crew, maintenance, and insurance to another airline or business entity (in this case IAG), which pays at a rate based on the number of hours operated and also covers the cost of fuel, airport fees and other duties. Dedicated freighter services give IAG Cargo the opportunity to service destinations that are not available on their passenger route network.

Airfreight the world over makes use of unit load devices (ULDs). These are aircraft containers or pallets that allow the assembly and consolidation of individual pieces of baggage, cargo and mail in order to facilitate rapid aircraft loading and unloading and to ensure efficient handling on the ground and in terminals. Used on both wide- and narrow-body aircraft, ULDs perform a critical safety function during flight by keeping cargo secure.

ULD pallets are flat sheets of aluminum with rims designed to lock on to cargo net lugs. They weigh between 3 and 6 tonnes each. Shipment

pieces are stacked on the base and then usually covered with a net to secure loose pieces.

ULD containers, also known as 'cans' and 'pods', are typically closed containers made of aluminum weighing between 1.1 and 1.5 tonnes. Some have built-in refrigeration units to conserve perishable cargo such as fruit, vegetables, flowers and fish.

Aircraft loads departing from Heathrow consist of containers, pallets, or a mix of ULD types, depending on requirements. The use of ULDs allows a large quantity of cargo to be bundled into a single unit. Since this leads to fewer units to load, it saves ground crews time and effort and helps prevent delayed flights. Each ULD has its own packing list (or manifest) so that its contents can be tracked.

There are several common types of contoured ULDs that are curved to fit different aircraft

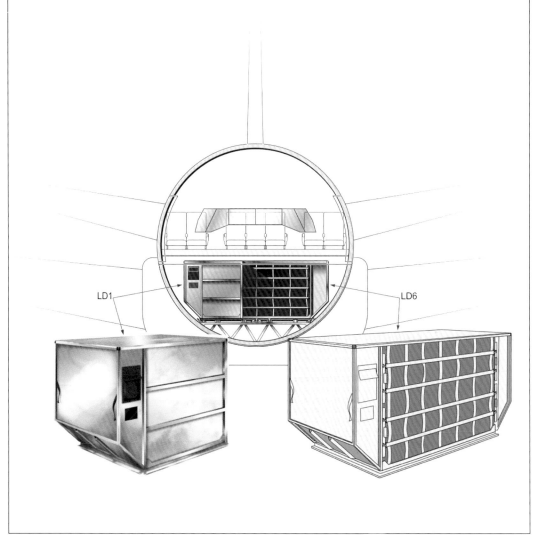

LD1   LD6

ABOVE **Crews use roof-mounted forklifts to prepare a cargo consignment on the 'break and build' floor.** *(Author)*

and thereby provide as much cargo volume as possible given the shape of an aircraft belly.

All ULDs are identified by their ULD number. A three-letter prefix identifies its type and how it should be moved, followed by a four- or five-digit serial number to uniquely identify it from others of the same type, and ending with a two-character (alpha-numerical) suffix identifying the ULD's owner; if an airline, this is often the same as its IATA designator code. For example, AKN 15902 BA means the ULD can be raised with a forklift and belongs to British Airways.

Cargo operations

BELOW **Schematic of the IAG Cargo centre at Heathrow.** *(Roy Scorer)*

The IAG Cargo operation is no ordinary business. It is one of the world's most sophisticated logistics operations, using state-of-the-art technology to deliver a world-class service.

In fact, 75% of the building is unmanned and thus reliant on automated procedures.

The main cargo building, which spans the size of six football pitches, is divided into two distinct areas:

■ Landside – outside the building there are 24 receiving doors: 2 are for high-security items, 8 for intact containers, and a further 10 for loose cargo where exporters and importers, either themselves or through their agents, deliver or collect shipments. Inside the building on the landside is the massive consignment store with cages for storage, a sorting area, and a large build-and-break area – freight arrives at the facility either in a state whereby it can be sent on as an intact unit or it may need to be broken down in such a way that it can be passed on to the customer.

■ Airside – the interior of the airside portion of the building provides storage capacity for more than 1,800 ULD containers, while the external area is where loads are moved to or from aircraft. Around 1,500 containers are moved to and from the facility every day.

Inside the building it's immediately clear that this is very different to a traditional warehouse. Once accepted, consignments are allocated to one of 8,000 dedicated cage storage locations, each of which has a unique bar code used to instantly track the item at any point on its journey. The cages also have weight sensors that help to identify loads that exceed the maximum weight allowed.

Each cage sits adjacent to part of the 21km

Consignment Store

IAGCargo

Office   Sortation Floor   Break & Building Floor   Dolly Dock   ULD Store

LEFT **Inside the IAG consignment store.** *(Author)*

of rail track that runs over five floors linked by two major lift shafts. Running on the track are 72 elevated cage transport vehicles that move up to 250,000 individual consignments through the building every month. It's easy to understand the reason for such a high number of movements when one considers that consignments only have a typical storage time of 6–24 hours in the facility. The technology in use in the building is so sophisticated it evenly balances the freight so as not to place any unnecessary loadings on the facility's structure.

Depending on the type of cargo being carried, the operations team must ensure the logistics chain is maintained throughout the journey. A good example of this is pharmaceutical materials requiring temperature control throughout the course of their delivery.

Cages with individual consignments headed to the same destination are automatically picked from their cage locations by the transport vehicles and brought down to the pre-flight assembly area. In one of 19 different areas the consignments are built for the specific flight, either as a container or a pallet. To keep the floor area clear and enhance safety, the build teams use roof-mounted forklifts to prepare the loads.

Using a hand scanner the bar codes for each consignment that make up any one build are scanned, and once recorded the job is closed, a final tag is printed with the final overall weight, and this is then added to the load control information prepared for the aircraft destined to carry the consignment. Just as passengers fly with some form of a ticket, all freight is accompanied by an air waybill (AWB). The AWB is a consolidation of information related to a specific shipment and includes the destination, origin and contents.

LEFT **Incoming cargo is transferred from a ULD container to one of the cages.** *(IAG)*

RIGHT Ground crews prepare to unload cargo from an Air China Boeing 777-300ER. *(Author)*

So efficient is the operation that the loads can be worked on up until 2½ hours before the flight is set to depart. In many cases the relevant cages are brought down in advance and buffered until ready, effectively buying the operations team valuable time.

Once a ULD is ready it is placed on one of more than 100 freight vehicles and ULD carriers (dollies) that are manned 24 hours a day, seven days a week. A team of 400 drivers working shifts then delivers the freight to the aircraft using a dedicated access road from the facility to Terminal 5, as well as the airport's cargo tunnel. The 885m tunnel is not open to the public – it is used only by vehicles

BELOW Cargo dollies lie in wait for the next arrival. *(Waldo van der Waal)*

RIGHT Sophisticated computer systems track the location of each cargo transporter vehicle across the airport. (Author)

with security clearance to drive airside. Once at the aircraft the cargo is loaded into the aircraft's belly using a high-loader.

Back in the operations centre, the airside planning department tasks drivers to specific jobs and tracks the location of each vehicle using a GPS. Freight must be transported to the aircraft at least 30 minutes before push-back. Drivers use handheld radio data terminals (RDTs), which provide them with the latest job information and allow for communications with the planning department.

In the operations centre, planners review the individual build status of each shipment on large computer monitors: green-coloured builds are progressing well, red-coloured builds are running behind and need urgent action. 'The biggest challenges that we face as a business are often those that are outside of our control, such as the weather and aircraft delays', says Mat Burton, General Manager UK Operations.

The IAG Cargo operation makes use of sophisticated IT systems that allow it to perform a wide range of tasks, from managing capacity and inventory to actually moving the freight and tracking its location to billing customers. The

UNUSUAL CARGO

The IAG Cargo team facilitates the movement of a hugely diverse range of cargo, from life-saving drugs to fresh fruit and vegetables to automotive parts. More unusual cargo includes equipment for the NASA space station, spare aircraft engines, wind turbine components, dinosaur fossils, replacement parts for oil rigs and the latest in high street fashions.

IAG Cargo has transported live and exotic animals including lions, rhinos, racehorses, kangaroos and even sharks. 'We work closely with the shippers to ensure that their animals have as comfortable a flight as possible, a point of care in which we take great pride,' says IAG Cargo's Global PR Manager, Adam Chaudhri. 'For example, when we transported Lola the hippo in 2012 we made sure that the shipper was able to stay with her until the last possible moment, thereby helping to reduce stress. We also gave her a shower just before the flight to ensure maximum comfort.

'With the amount of fresh flowers we freight, I think it's safe to say IAG Cargo delivers Valentine's Day to the UK,' he adds.

The operation employs a special team dedicated to working with 'asset protection' of high-value goods, as IAG's aircraft regularly carry a wide range of valuable cargo including precious artworks – such as the Terracotta Warriors – and supercars. 'In 2013, for example, we successfully transported a twin turbo 720hp Pagani Huayra, valued at £850,000, from Frankfurt to Hong Kong on our Boeing 747-8F,' says Chaudhri. It seems there will always

be a demand for air cargo to transport high-value, time-sensitive shipments.

Furthermore, the operation regularly offers cargo assistance in the event of a humanitarian crisis. For instance, when the Haiti earthquake struck in 2010 a Boeing 747 was cleared from its schedule and used to transport 50 tonnes of relief aid to the stricken country. As well as placing cargo in the hold, vital supplies including medicines, tents, water purification systems and kitchen equipment were transported in the cabin by the volunteer crew.

BELOW Cargo comes in all shapes and sizes. (IAG)

movement of the ULDs and cages is supervised from a central control room with cameras which pan 360° and zoom, giving controllers full visibility of the multitude of automated processes taking place in the building.

When it comes to actually loading the aircraft, the team commences with the equivalent of a massive 3D jigsaw puzzle in the hold, using

THE LOCKERBIE DISASTER

The devastating threat posed by explosives in the hold was evidenced during the catastrophic bombing of a Boeing 747 – Pan Am 103 – over Lockerbie in Scotland in December 1988. An explosive device placed in the forward hold detonated, killing all 243 passengers and 16 crew members. Large sections of the plane crashed into Lockerbie, in southern Scotland, killing a further 11 people on the ground. The Air Accidents Investigation Branch (AAIB) concluded that the explosion punched a 20in hole on the left side of the fuselage and that the nose of the aircraft separated from the main section within three seconds of the explosion. Although the explosion was in the aircraft hold the effect was magnified by the large difference in pressure between the aircraft's interior and exterior.

At the time of the disaster airport security was certainly more relaxed than it is today. Although passengers boarding flights were required to pass through metal detectors and have their hand baggage examined for the presence of concealed weapons, their hold baggage underwent few security controls.

every available inch to secure the maximum possible payload.

Types of cargo

A host of different cargo options are available, each of which is created to address a specific need:

- **Priority** – used when speed and reliability are of the essence; this freight enjoys preferential status at all times.
- **Constant Climate** – a precision, temperature-controlled service used for temperature-sensitive pharmaceutical materials.
- **Airmail** – designed to meet the specialised requirements of the world's postal operators.
- **Perform Loose** – a general cargo service for loose cargo requirements.
- **Perform Unitised** – a general cargo service for a range of unitised cargo requirements.
- **Courier** – designed around quality and speed of deliveries of urgent cargo.
- **Dangerous Goods** – IAG Cargo transports dangerous goods around the world in the safest possible way, under the strictest regulations.
- **Human Remains** – a caring and compassionate service ensuring that human remains are treated with the utmost dignity and are brought home with the absolute minimum delay.
- **Secure** – high-security service is used for valuable consignments, with relevant staff and customers required to meet strict security vetting.
- **Live Animals** – the shipment of live animals is handled by dedicated personnel with vast experience in this area.
- **Constant Fresh** – designed to keep perishable goods fresh and thereby optimise their shelf life.

Security

Although passenger screening often gets the spotlight, the process of overseeing efficient cargo operations while maintaining tight security procedures is a vitally important task at Heathrow. The cargo facility is a high-activity area dealing in vast volumes of freight and the security process would be much simpler if there were cargo-only flights, but today virtually every passenger flight departing from Heathrow carries some freight.

The technology used for cargo screening

has to be considerably larger than that used on passenger baggage systems, but performs essentially the same function. The equipment has to be large enough to scan palletised cargo for suspicious items, and staff need to be carefully trained to accurately detect what are smaller needles in much bigger haystacks compared to traditional cabin baggage. At Heathrow large-scale X-ray machines are operated by a third-party supplier that specialises in this field, with scanning requirements and procedures set by the DfT.

Similar scanning technology is used to provide security checks on the supplies brought on to the airport's premises, such as in-flight catering. Heathrow also operates a substantial duty-free retail operation, which together with the vast array of food and beverage outlets across the five terminals translates into hundreds of tonnes of food, drink and goods that need to be checked. In some instances secure supply chains are in place that go all the way back to the initial point of production and distribution, with secure escort services bringing the supplies to the airport.

The Heathrow Animal Reception Centre (HARC)

Situated nearby and working in close conjunction with the Cargo Centre is the Heathrow Animal Reception Centre (HARC). Formerly known as the Animal Quarantine Station, the HARC has established itself over the past 25 years as a world leader in the care of animals during transport.

The centre is effectively the live animal border inspection post at Heathrow. All animals entering the EU from outside the European Union have to pass through such a post so they can be inspected and have veterinary documentation issued.

A dedicated team of 30 staff run the centre, which is open 24 hours a day, 365 days a year. The team receives and cares for more than 80 million animals each year, including 45 million invertebrates, 7 million live eggs, 28 million fish and some 13,000 cats and dogs. It's not uncommon for the centre to regularly play host to tarantulas, cobras, prized racehorses, tigers, cattle and even baby elephants.

Animals arrive at the centre as a consequence of being in transit either through or to Heathrow Airport from all over the world. Once inspected and having cleared customs, animals are reunited with their owners.

HEATHROW PEOPLE: MAT BURTON

General Manager UK Operations IAG Cargo

My working day at Heathrow starts with a daily ops meeting that provides an in-depth performance review of our two facilities at the airport, Ascentis and Premia. We also look across the next 48 hours to spot any issues that may impact the operation.

Items we need to keep an eye on can often be cargo that is out of the ordinary, and these can vary greatly – some weeks we will be carrying race cars for international competitions, and other weeks we will be moving spare aircraft engines to our bases around the world. In fact, this morning sees the arrival of two rather exotic racing cars that have been flown in especially for the Le Mans 24-hour race in France. We also carry a large amount of relief aid, which can require very quick responses from our teams when required.

Our facilities are extensive, and are continually undergoing upgrades and improvements, such as our new Constant Climate Centre. This is a fully chilled facility that will improve the capacity we have for temperature-sensitive goods such as pharmaceuticals. Our planning horizon is based on the next 10–15 years, so it's vital we understand the impact of our planning and ensure we're taking the right decisions.

With a cargo operation that effectively operates 24/7, 365 days a year, there's no set time to head home, but before I do I call round to the key members of the team to see how the day has gone. It is vital that we are aware of even the smallest issue, because without checks in place those problems can multiply.

I started work with British Airways in 1983 and my career has taken me to far-flung corners of the world and today I am responsible for overseeing and enhancing the performance of our UK operations.

CHAPTER 11

The future of Heathrow

OPPOSITE A Malaysian Airlines Boeing 747-400 comes in to land on runway 27L. Today Heathrow sits at the heart of an international transportation crossroads. *(HAL)*

Introduction

Flying is of great value to the United Kingdom, for the economy, for society and for consumers. It fosters investment and trade and links multicultural Britain to an increasingly globalised world. What matters most to travellers is being able to get where they want to go, when they want to go. Heathrow's strong network of short-haul and long-haul traffic enables it to offer a wide-range of destinations that 'point-to-point' UK airports cannot match. Heathrow is able to serve important long-haul destinations at higher frequencies with bigger aircraft, which benefits London and the UK.

However, unlike airlines which are able to move relatively quickly to respond to changes in traffic flows by leasing or retiring aircraft capacity, airports must make long-term planning decisions that can sometimes run 50 years into the future. Heathrow is no exception, particularly as it is the UK's only international hub airport and a vital piece of national infrastructure. As the preceding chapters have shown, Heathrow has developed from somewhat humble beginnings described in the opening chapter to become the world's busiest airport by international passenger traffic. But a lack of space to expand is allowing rivals in the Middle East and mainland Europe to increase their airline networks and passenger capacities and Heathrow is struggling to remain competitive.

AVIATION'S CONTRIBUTION TO THE ECONOMY

- The UK has the second-largest aircraft manufacturing industry in the world.
- Aviation benefits the UK economy through its direct contribution to gross domestic product (GDP) and employment, and by facilitating trade and investment, manufacturing supply chains, skills development and tourism.
- The aviation sector contributes around £18 billion per annum of economic output to the UK economy and generates £53 billion in annual turnover. It employs around 220,000 workers directly and supports many more.
- The UK has the third-largest aviation network in the world, after the USA and China.
- The number of passengers using non-London airports has increased by over a third since 2000.
- London offers at least weekly connections to over 190 destinations. In comparison, Paris serves around 300 destinations and Frankfurt around 250.

Source: Aviation Policy Framework Document.

Heathrow operates in a competitive environment, pitching for customers with other European hubs such as Amsterdam's Schiphol airport, Charles de Gaulle in Paris and Frankfurt in Germany. In a burgeoning aviation sector – where there is compelling evidence for continued growth in demand at Heathrow – the airport and its airline customers promote a shared vision to make the airport Europe's 'hub of choice'. Boeing, for example, estimates that the global large commercial aircraft fleet will double by the year 2031, with many of these aircraft sure to be landing at some point at Heathrow. Heathrow believes the most important way to cope with this demand and deliver the vision is to prioritise continuous improvements in the 'passenger experience'.

Over the longer term, this ultimately means investing in two fundamental areas – the airport's infrastructure (better terminals, facilities and access to the airport) and capacity (additional runways to reduce current pressure on the airport and more travel choices for passengers). Any new initiatives have to be carried out in an affordable manner, are sometimes fraught with political and social challenges, and need to take into consideration the impact of climate change and noise pollution while at the same time keeping charges levied on airlines within a competitive range.

Whatever the future holds for Heathrow it continues to develop apace, with innovative ideas, projects and infrastructure to ensure that it maintains its position as one of the world's leading airports. This chapter explores what lies in store for Heathrow in the years to come.

Forecasting future demand

One of the key determining factors that impacts almost all aspects of the airport's future planning and development is the future traffic forecast. Heathrow works collaboratively with the airlines to jointly create and refine its passenger forecast. The aim of forecasting is to determine how patterns of demand will alter over time, reflecting external factors such as growth in incomes, changes in ticket prices and demographic variations.

Heathrow passenger volumes grew steadily through the 1990s, reaching 64.3 million passengers in 2000. Recession and the terrorist attacks in September 2001 led to a sharp fall in volumes early in the last decade, with some

recovery through to 2007 as the world economy grew. Although traffic fell between 2008 and 2010, it grew to its highest ever level in 2013, when the airport recorded a total of 72.3 million passengers. During this time, passenger traffic to the vital growth markets of India, Brazil and China all saw sizeable increases.

However, with Heathrow now at more than 99% of its 480,000 air traffic movements constraint, opportunities for traffic growth are limited. DfT forecasts indicate that there will be an unserved passenger demand of 11 million at Heathrow by 2020 and 28 million by 2030.

With this in mind there is clearly compelling evidence for continued growth in demand to travel through Heathrow. The airport's exposure to global markets, including emerging economies with higher potential for increased levels of flying as they grow richer, also supports the case for future demand growth.

The case for a single UK hub

As an outward-looking nation with an island economy that for centuries has owed its prosperity to the transport and trade routes linking it with the rest of the world, much of the UK's future prosperity will continue to be shaped by the effectiveness of its international transport networks, and Heathrow lies at the very centre of this hypothesis.

Unless something is urgently done to address the issue, ten years from now Heathrow's operating capacity will have been constrained for two decades, and all the while foreign governments, airlines and hub airports such as Dubai and Istanbul will have continued to make major investments that exploit the UK's hub capacity constraint. Heathrow recognises that the current political and planning landscape means it will likely be 2026 before any significant additional hub capacity could be operational in the UK, though the airport's management team believe it is at Heathrow that this can be delivered the quickest.

There have been suggestions of creating a dual hub between Heathrow and Gatwick, but history and international experience show that this failed in the past as airlines were ultimately attracted back to Heathrow, where they could maximise transfer opportunities.

As the country's only international hub airport, Heathrow supports flights that cannot be operated profitably from any other UK airport and connects British businesses with the growth markets other non-hub airports cannot reach. Heathrow serves 75 direct destinations worldwide that are not served by any other UK airports, and handles more than 80% of all long-haul passengers that come to the UK.

To be competitive, hubs need to be able to attract network airlines and their passengers. Airlines will compete with each other and will move operations to hubs that improve their profitability.

ABOVE The airport has had to prepare itself for a significant increase in bigger aircraft such as this Singapore Airlines A380. *(HAL)*

THE AIRPORTS COMMISSION

At the end of 2012, while the importance of hub airport capacity was beginning to be accepted by the UK government, it established the independent Airports Commission, tasked with examining the need for additional UK airport capacity. The Commission received 75 proposals and an interim report published in December 2013 saw two potential sites selected for further assessment.

Heathrow
- A new 3,500m runway constructed to the northwest of the existing airport, as proposed by Heathrow Airport Ltd and spaced sufficiently to permit fully independent operation.
- An extension of the existing northern runway to the west, as proposed by Heathrow Hub Ltd, lengthening it to at least 6,000m and enabling it to be operated as two separate runways – one for departures and one for arrivals.

Gatwick
- Analysis will be based on a new runway over 3,000m in length spaced sufficiently south of existing runway to permit fully independent operation.

A final report will follow in the summer of 2015 (after the next general election) and will assess the environmental, economic and social costs and benefits of various solutions.

This competition is good for consumers, delivering lower prices and greater choice of services.

A hub airport is an airport where local passengers combine with transfer passengers to allow airlines to fly to more destinations more frequently than could be supported by local demand alone. Put simply, it is the most efficient way of connecting many different destinations. Typically, passengers from short-haul flights combine with passengers from the airport's catchment area to fill long-haul aircraft. Transfer passengers are essential for a hub airport to serve many destinations. They allow the UK to connect to countries where it couldn't sustain a direct daily flight itself.

For example, London, Edinburgh or Stockholm might not have enough people wanting to travel to São Paulo to be able to justify a daily flight. By pooling demand through a hub airport like Heathrow, airlines are able to serve the destination profitably and increase the number of flights per day. This hub phenomenon is self-sustaining – once a flight to São Paulo has been established, then more passengers travel through the hub and more passengers are available to transfer on to other flights.

The reason Heathrow is able to support direct long-haul flights to multiple destinations is because of transfer passengers. One-third of passengers on the average flight from Heathrow are transfer passengers, and the vast majority of flights have at least 25% transfer passengers. If these passengers did not exist then the routes would not be viable and would either disappear or diminish in frequency.

Ultimately these passengers and flights support trade, jobs and economic growth. Without an extensive array of connections through a national hub it is estimated that the cost to the UK economy could already be up to £14 billion a year, and this figure could rise to up to £26 billion a year by 2030. This leaves the government with the challenging choice of either ignoring the issue, knowing that this is likely to result in the UK falling behind its European competitors at the cost of lost growth and jobs; closing Heathrow and replacing it – at a massive cost – with a new hub airport altogether; or finding a solution that will increase capacity at Heathrow.

Runway capacity

Studies show that by 2021 gross domestic product in Britain will drop by £8.5 billion per year due to an over-stretched Heathrow lagging behind global aviation growth. But it is not just Heathrow that is under pressure. DfT figures indicate that all of the airports serving London will be full by 2030, perhaps even as early as 2025 if no new runways are built. To address the capacity crunch in the south-east of England, there are really only three options for consideration, and Heathrow obviously lies at the very heart of the debate:

- To expand Heathrow into a 'super hub'.
- To expand one of London's other airports (Gatwick, Luton or Stansted).
- To build an entirely new hub airport.

Whatever the solution, the only way of creating a meaningful increase in flights and routes is to add additional runway capacity, and that is seen as a political 'hot potato', with government ministers very reluctant to make a decision that could bring extra noise to hundreds of thousands of city dwellers.

Wind the clock back to 2009 and the government had approved a third runway at Heathrow. It was set to be 2,200m long and located to the north of Heathrow, at Sipson. The new runway would theoretically allow for an increase in the number of permitted movements from

480,000 to 605,000 and would cater for short-haul domestic flights, thereby freeing-up space on the main runways for more, bigger aircraft. Ultimately a third runway would provide some much-needed extra capacity, but despite some light at the end of the tunnel for Heathrow, a bitter blow came in May 2010 when the (new) government withdrew support for a third runway, raising the question of whether airlines could commercially pursue the same investment and growth path at an indefinitely constrained two-runway Heathrow.

Since then the government has announced the formation of the Airports Commission to which Heathrow submitted three proposals for a third runway on the basis that:

■ Each solution was quicker and cheaper than building a new hub airport.
■ All solutions have fewer people affected by noise than Heathrow today.

■ Three runways are enough to maintain the UK's global hub status for the foreseeable future.

The official Heathrow proposals included a third runway placed to the north, north-west or south-west of the existing airport. The airport believes all three options can deliver the required extra capacity by 2025–29 at a cost of £14–18 billion. Heathrow believes the two westerly options offer clear advantages by delivering a full-length third runway while minimising the impact on the local community from noise and compulsory house purchases.

Each westerly option would raise the capacity at Heathrow to 740,000 flights a year (from the current limit of 480,000). That would cater for 130 million passengers, allow the UK to compete with our international rivals and provide capacity at the UK's hub airport for the foreseeable future.

On the day of the Heathrow announcement,

LEFT The debate about airport capacity in the United Kingdom has dominated recent newspaper headlines. (Author)

Heathrow's own initial proposals included a third runway to the north-west of the airport (above). At 3,500m it provides maximum capacity, flexibility and resilience. Another option (below), rejected by the Airports Commission, proposed a third runway to the south-west that would take longer and cost more to build. *(HAL)*

Heathrow's chief executive, Colin Matthews, said, 'After half a century of vigorous debate but little action, it is clear the UK desperately needs a single hub airport with the capacity to provide the links to emerging economies which can boost UK jobs, GDP and trade. It is clear that the best solution for taxpayers, passengers and business is to build on the strength we already have at Heathrow.'

A new Heathrow would benefit from already planned public transport improvements, such as Crossrail, Western Rail Access and High Speed 2, and the charges per passenger would be likely to be much lower than at a new hub airport. And despite the increase in capacity, the total number of people affected by noise from aircraft will fall. This is due in part to the westerly options being positioned further from London than the existing runways. Each mile the runway is moved to the west puts arriving aircraft approximately 300ft higher over London. Continued improvements in aircraft and air traffic technology will also result in fewer people being disturbed. As a result, even with a third runway there will be 10–20% fewer people within Heathrow's noise footprint in 2030 than today.

Expansion at Heathrow can also be met within EU climate change targets. This is made possible by continued improvements to aircraft efficiency which mean that air traffic could double by 2050 without a substantial increase in emissions. If carbon trading is included, emissions would be reduced. Similarly, Heathrow would improve local air quality in line with EU standards because of cleaner vehicles and the increased proportion of passengers using public transport.

In addition to the official proposals from Heathrow, the Airports Commission received more than 75 submissions to address the country's aviation capacity issues. Most have since been rejected, but some of the more interesting ideas included:

■ Economist Tim Leunig put forward a proposal to build four new parallel runways at Heathrow, arranged in two sets of pairs, immediately to the west of the existing airport. The idea had some significant implications for surrounding areas, not least of which would have seen the M25 motorway tunnelled underneath the runways. Moving the runways west would have reduced noise levels over west London, since the planes would be higher over any given place. Leunig further proposed the airport reuse all existing terminals except Terminal 4. All existing

Comparison of options at Heathrow

	Heathrow today	North option	North-west option	South-west option
Passenger capacity	80 million	123 million	130 million	130 million
Maximum flights	480,000	702,000	740,000	740,000
Cost	–	£14 billion	£17 billion	£18 billion
Length of new runway	–	2,800m	3,500m	3,500m
Noise (population within the 57dBA Leq contour)	243,000	-10%	-15%	-20%
Residential properties lost	–	2,700	950	850
Opening date	–	2025	2026	2029
Ecology impact (hectares)	0	0	0	716
Volume of flood zone 3 storage lost	–	6,000m³	116,000m³	1,416,000m³
Grade I/II listed buildings lost	–	0	2	0
Construction complexity	–	Low	Medium	High

air traffic control, refuelling lines, maintenance and engineering facilities would remain. The Heathrow Express, Crossrail and Piccadilly Lines would continue to serve the airport, and would be extended to reach the new terminal. 'The airport would have twice the capacity of the current Heathrow, implying a maximum of around 960,000 movements and 140 million passengers. A sensible working maximum would be 850,000 movements, and 121 million passengers,' said Leunig.

■ Another option for Heathrow, devised by Concorde's longest-serving pilot, William Lowe, proposed to extend Heathrow's existing runways to 7,500m, moving part of the M25 in the process. The plan to lengthen each runway would accommodate aircraft landing and taking off at the same time, using the mixed mode of operations. The plan estimated to double passenger capacity from 70 million to 140 million and increase the number of flights from 480,000 per year to almost a million.

■ Consideration was given over many years to expansion at London's other airports to deal with broader capacity issues in the south-east of England. Gatwick has demonstrated just how well a single-runway airport can service a large volume of flights. However, for a national hub to be successful it is felt that a three- or four-runway airport is necessary, and it is hard to see that happening at Gatwick and legal agreements prevent the building of a new runway at Gatwick before 2019 at the earliest.

■ Luton in Bedfordshire is well located and is

close to a high-quality rail line to London. There is also land available to the east and south of the airport for expansion, but the facility is not very well configured to be a major international airport – it has only one short runway and has grown in an ad hoc way over time, with a range of operational challenges. Stansted has capacity but has limited transport links to London.

■ The only other alternative to expansion at Heathrow (or any of London's other airports for that matter) is the creation of an entirely new hub, more than likely to the east of London in the Thames Estuary. The 'London Britannia Airport' proposal claims to generate 200,000 jobs but has been slated by environmental groups and at £47 billion, pro-Heathrow commentators argue that the new hub would simply cost too much. Critics further argue that opening an airport east of London would mean not only the inevitable closure of Heathrow but might also create a significant threat to City Airport. They argue further that redeveloping west London without Heathrow would take decades, causing unparalleled social and economic upheaval. A further concern is that an airport located this far out of the city would be difficult to reach – a key part of Heathrow's success has been its location to the west of London, and its proximity to the M4 corridor, where some of the world's largest multinational companies are headquartered.

BELOW Traffic from across the Atlantic is likely to continue increasing at a steady rate. *(HAL)*

Airspace issues

It is not only UK runway capacity that is constrained – so is the country's broader airspace system, which is amongst the most congested in the world. This is perhaps not unsurprising given that the UK's current airspace system – some of the most complex in the world – has not been significantly updated for more than 40 years. Airspace is a key part of the country's transport infrastructure and is vital to all airspace users, from airlines to the military and private fliers.

Aviation authorities including the CAA, NATS, DfT and Ministry of Defence (MoD) are at the forefront of a range of measures to increase airspace capacity, improve flight efficiency, and reduce aviation's environmental impact. The Future Airspace Strategy (FAS) includes various measures to address airspace capacity, many of which would have a direct bearing on Heathrow, including:

■ Allowing aircraft continuous climb-outs on take-off that get them to their optimum cruising altitude as quickly as possible.
■ Providing aircraft with more efficient routings that save time and fuel.
■ Better management of arrivals at airports, such as reducing the time that aircraft are in holding stacks.
■ Linking the whole aviation network together to share up-to-date flight information, thereby enabling better operational decisions and increasing resilience to unexpected events.
■ Using the latest technology throughout the system to increase airspace capacity and safety.

NATS, for example, have consulted on a change to the airspace in Terminal Control North (a wide area covering North London and parts of East Anglia). This proposes changes to holding patterns and arrival and departure routes for Heathrow and other airports in the area, in particular to take account of precision navigation, the need to reduce holding and distance flown and allowing for traffic growth. Ultimately, any measures must ensure that safety standards are maintained and can accommodate the forecasted increase in air transport movements. Once the strategy has been agreed it will be tied into wider projects such as the European Commission's 'Single European Sky' initiative, which aims to streamline the way airspace is used across the Continent.

Airport Collaborative Decision Making

An interesting development in the international aviation industry that will improve efficiency levels at Heathrow is the introduction of Airport Collaborative Decision Making (A-CDM). Testing and implementation of the system is already under way at Heathrow.

A-CDM supports the flow of information (and hence subsequent actions) from the time an aircraft lands until it takes off again. An accurate estimate of the time an aircraft is due to land enables a stand to be assigned confidently in good time, which allows the airline staff and ground handlers to be ready to meet passengers off that aircraft and for the turnround process to begin promptly.

As long as an aircraft expects to be ready to depart at its scheduled time there is no need to amend any times during the turnround process, but if it is thought that the aircraft will be ready to leave earlier or later than planned then the time when it will be ready must be updated and shared. This allows events further down the line to be planned more accurately.

As this information is shared with all stakeholders, it allows airline staff and ground handlers to prioritise their work to ensure the flight does depart as planned and allows those responsible for managing stands the relevant information required to allocate the newly vacated stand to the next arriving aircraft – thus starting the process again.

A live map of the airport shows exactly where all arriving and departing aircraft are on the airfield, providing a quick and easy way of getting a high-level impression of the situation.

The system has already shown positive results in other European airports, with savings equating to significant reductions in both fuel usage and carbon emissions.

Accessibility

The future will see Heathrow continue to devise and implement transportation strategies with a view to improving the passenger experience and placing the airport at the centre of the national rail network. This will include shifting both passengers and employees from car to rail, reducing emissions and the impact of road congestion.

A new programme, called the Wider Heathrow Integrated Rail Strategy (WHIRS), seeks to build on previous investment by ensuring that Heathrow has fast, frequent and comfortable rail connections for passengers, while at the same time significantly improving links to the surrounding community. Priorities include:

- Ensuring that Crossrail provides passenger-friendly, convenient connections for Heathrow travellers.
- Continued enhancements to the Heathrow Express service.
- Consideration for rail access from the west of Heathrow, providing a direct connection with Slough, Reading and the Thames Valley.

- Further discussion on how to connect the airport to the south of the country.
- Using the strategic nature of Heathrow as a UK transport node and its ability to act as an interchange and 'hub' for bus, coach and rail routes.
- Exploiting the fact that Heathrow will be served by a spur from the main London–West Midlands high-speed line.

Major development projects

Heathrow is constantly working on a range of projects to build and improve facilities, including large-scale capital investment programmes that will collectively deliver both an enhanced experience for passengers and a more competitive hub airport for airline customers. Some of these include:

- The current centrepiece of Heathrow's infrastructure transformation is the construction of the new Terminal 2. Once open, more than 60% of passengers using Heathrow will enjoy some of the newest airport facilities in the world. Terminal 2 will be very similar to Terminal 5, as Heathrow's owners set about rebuilding the airport in what some describe as a 'toast rack' format of neat parallel rows of terminal and satellite buildings.
- Within other terminals there is significant restoration and modernisation, with 'back-office' upgrades for staff, new retail opportunities and additional security lanes for passengers.
- An integrated baggage system will replace the baggage infrastructure in Terminal 3, improving minimum connect times and miss-connect rates.
- To optimise capacity within the constraint of 480,000 air traffic movements, the airport has embarked on a programme to deliver changes that will improve the resilience of operations, by improving punctuality, predictability, and the ability to reorganise runway usage during periods of unplanned high demand.
- Various new IT systems for security, radio and cellular infrastructure and baggage systems will reduce operating costs and deliver improved value.
- Following the snow disruption in 2010, further investment will be made in equipment to support winter resilience, including snow-clearing equipment, additional glycol storage facilities, snow-melting equipment and an improved command and control centre.

Dealing with future aircraft types

Heathrow is already capable of dealing with 'Code F' aircraft (one of six generic ICAO categories designated to the biggest commercial aircraft based on their wingspan and main landing gear width), but as more airlines increase their use of bigger aircraft, such as the A380, Heathrow will need to further adapt its infrastructure to meet these ongoing needs.

Between 2003 and 2014 Heathrow has invested some £11 billion in new facilities that have improved operational performance and customer service, many of which are linked to the advent of new aircraft types. Some of these investments have included the widening and strengthening of taxiway pavement and stands, new stand entry guidance systems, new air bridge configurations and changes in the terminal to deal with the increased numbers of passengers, as well as the acquisition of new fire equipment capable of reaching the upper deck of the super jumbo.

Sustainability

Delivering a sustainable airport is one of Heathrow's key future goals, and underpins the vision for Heathrow to be 'Europe's hub of choice'. This requires Heathrow to find the right balance between economic, social and environmental objectives, enhancing the positive impacts the airport brings, while minimising the negative impacts and meeting agreed environmental limits. The goals include:

- Climate change: by 2020 reducing carbon emissions from energy use in fixed assets at the airport by 34% compared to 1990 levels.
- Air quality: delivering full compliance with EU air-quality limits.
- Waste: by 2020 recycling 70% of airport waste.
- Noise: limiting and, where possible, reducing the impact of noise at the airport – new, quieter aircraft such as the Boeing 787 have contributed towards ensuring that fewer people are affected by noise from Heathrow today than at any time since the 1970s, even

though the number of flights has almost doubled. Despite their size and impressive 60m wingspan, Dreamliner 787s feature a revolutionary design that allows for a noise footprint 60% quieter than that of similarly sized aircraft. By 2020 it is expected that approximately 30 Airbus A380s and 60 B787s will be using Heathrow. An A380 carries roughly 42% more passengers yet produces half the noise when taking off compared to a Boeing 747-400, the iconic long-haul airliner for the last 30 years.

A DAY IN THE LIFE OF MATT GORMAN

Heathrow Sustainability Director

I live to the west of the airport and there's not yet an easy way to get to work by train, but there will be from around 2020, when a 'western rail link' is built to link the airport directly to Slough, Reading and the west. Until then I take advantage of Heathrow's car share scheme and share my journey with a colleague who lives close by.

My day starts with a monthly 'Environmental Management Committee'. This brings together all of the key staff from our operations teams in the airport terminals and airside to review how we're doing in improving environmental performance.

A major focus at the moment is waste recycling. Heathrow is like a small city that generates lots of waste so by 2020 we're aiming to recycle around 70% of that waste.

I spend time chairing a meeting on Sustainable Aviation – a unique coalition which brings together all of the biggest companies in the UK aviation industry including airlines, aircraft manufacturers and other stakeholders like NATS. The group was formed in 2005 to develop and implement solutions for cleaner, quieter, smarter flying and a big focus at the moment is developing sustainable alternative fuels for aircraft.

I spend time with a group of government officials from China in the new Terminal 2 facility. We've incorporated some advanced design features to make the building as sustainable as possible, and governments and airports in other parts of the world are interested in what we've done. The building will be 20% powered by renewables – for example using woodchip – all sourced from within 100 miles of Heathrow. As well as being more environmentally friendly, this is also creating jobs in the local economy around Heathrow.

It's coming up for 16:00 and I head off to visit our airside team to see some of the new hydrogen vehicles we're trialling around the airport. This is a new technology that's much less polluting than traditional petrol or diesel vehicles. We make the hydrogen on site – all you need is water and electricity to make it, and when it fuels the vehicle the only thing it produces is a little bit of water, so it really is 'zero emission' technology. That's great for Heathrow, where the airport contributes to local pollution levels. We're committed to doing everything we can to cut pollution.

We know that while Heathrow brings lots of jobs and economic benefits locally, an airport is not always the quietest neighbour, and we're committed to doing all that we can to reduce noise. We've been discussing with communities close to the airport and in London how we can route aircraft differently at night to give people predictable breaks from the noise, and have been trialling some new operating procedures. Tonight we're sharing how that trial's worked and getting local feedback. When this is done I can call it a day!

Epilogue

By providing passengers with frequent and direct connections, millions of people rely on Heathrow every day to visit friends, take holidays, reunite families or simply get their business done. As such, Heathrow is one of the few places where the world meets – and in today's global economy, the airport's success is Britain's success.

The complex choreography of Heathrow's future is inextricably linked to key developments and trends in global aviation, which, amongst other things, will see increased competition from a raft of new airports. Heathrow has already made great strides to accommodate newer and larger aircraft coming into service, and this trend will only continue, requiring further investment, technology and space. Heathrow will also have to cater for expected growth in global passenger numbers – some industry experts believe the number of global travellers will double by 2030, with others forecasting some 16 billion annual passenger journeys by 2050.

Heathrow will also need to expand plans to deal with the way passengers buy travel services and use self-service technology along their journey. Linked to this is smarter and more efficient security technology, while intelligent urban planning and architectural design will see Heathrow become more efficient and sustainable.

Whatever the future holds, one thing is certain – Heathrow is a truly remarkable operation in every sense of the word, and its proximity to the capital city it serves, coupled with a devoted team of people, award-winning infrastructure and 'best-in-class' operational procedures, will see it firmly remain one of the world's most dynamic travel hubs for many years to come.

Whether it's an excited passenger heading off on holiday, a weary member of staff heading home or an enthusiastic author making one last visit to the airport to finish his book, Heathrow is certainly Europe's hub of choice, and does everything possible to live up to its mantra of 'making every journey better'.

(Author)

Acronyms and glossary

AAIB Air Accidents Investigation Branch.

AAU Apron Area Unit.

ACARS Aircraft Communications Addressing and Reporting System. A digital datalink system for transmission of short messages between aircraft and ground stations via radio or satellite.

A-CDM Airport Collaborative Decision Making.

ACL Airport Coordination Limited.

ACN Aircraft classification number.

ADIS Airport display information system.

ADM Airport Duty Manager or Airside Duty Manager.

AFS Airport Fire Service.

AGL Aeronautical ground light.

AP Aiming point. A point located forward of the centre of the touchdown zone positioned so as to compensate for the angle of approach from a pilot's point of view. Intended to be used as a visual reference by the pilot, it corresponds to where the glide slope intercepts the runway surface.

APM Automated people mover.

APOC Airport-wide Operations Centre.

Apron A defined area intended to accommodate aircraft during the loading or unloading of passengers, mail or cargo, and for refuelling, parking or maintenance.

APU Auxiliary power unit.

ASD Airside Safety Department.

ASDA Accelerate stop distance available.

A-SMGCS Advanced Surface Movement Guidance Control System.

ATC Air traffic control.

ATCO Air traffic controller.

ATIS Automatic terminal information service.

ATM Aerodrome traffic monitor.

ATSA Air traffic service assistant.

AWB Air waybill.

BA British Airways.

BAA British Airports Authority.

BAE British Airways Engineering.

Blast pads Concreted areas at the end of each runway designed to prevent erosion by aircraft running their engines up to full power before they start their take-off runs.

CAA Civil Aviation Authority.

CAS Controlled Airspace.

CCF Central Control Facility.

CDA Continuous descent approach.

CFMU Centralised Flow Management Unit, based in Brussels.

CGA Clear and graded area. An area within the runway strip where the ground is prepared so that if an aircraft runs off it will not sustain significant damage.

Cranford Agreement A verbal agreement made in the 1950s to avoid use of the northern runway for take-offs in an easterly direction over the village of Cranford.

CTA Central terminal area.

DFM Duty Field Manager.

DfT Department for Transport.

DMA Duty Manager Airside.

DME Distance measuring equipment.

EASA European Aviation Safety Agency.

EFPS Electronic flight progress strips.

EHM Engine Health Monitoring.

FEGP Fixed electrical ground power.

Fillet Widening of a taxiway at a junction or bend to accommodate an aircraft's main undercarriage as it turns.

FIR Flight Information Region.

FOD Foreign object debris.

Glide-path or **glide slope** A tall mast with three horizontal antennae, offset to the left-hand side of a landing runway. Part of the Instrument Landing System.

GMC Ground movement controller.

GMP Ground movement planner.

Go-around Procedure adopted when an aircraft on final approach aborts landing and climbs away from the airport.

GRE Ground run enclosure.

HARC Heathrow Animal Reception Centre.

HEART Heathrow Engineering Acquisition and Remote Telemetry system.

HMRC HM Revenue & Customs.

HOCC Heathrow operational conference calls.

HOEC Heathrow Operational Efficiency Cell.

Hold or **stack** Area of airspace in which aircraft approaching Heathrow can be held in an orbital pattern pending a landing slot becoming available.

IAG International Airlines Group.

IATA International Air Transport Association.

ICAO International Civil Aviation Organization.

ILS Instrument Landing System.

IRVR Instrumented Runway Visual Range.

kt Knots.

LACC London Area Control Centre.

Landing interval The amount of separation between aircraft on approach.

LARS Lower airspace radar service.

LATCC London Air Traffic Control Centre.

LDA Landing distance available.

Localiser Part of the Instrument Landing System at the very end of each runway. Resembles an orange 'toast rack' on poles.

LPO Lighting panel operator.

LRP Landing rate prediction. Calculation of the amount of traffic that should be able to land, dependent on aircraft type and weather conditions.

LTCC London Terminal Control Centre.

LTMA London Terminal Control Area.

LVPs Low-visibility procedures.

Manoeuvring area The part of the airport provided for aircraft to take off or land (the runway) and taxi to or from their parking areas (taxiways).

MARS Multi-aircraft ramp system.

MAU Manoeuvring Area Unit.

MCA Multi-choice apron.

Missed approach Another name for a 'go-around'.

MLW Maximum landing weight.

Movement Area Generic term for the apron area, the manoeuvring area, hangars and maintenance areas.

MTOW Maximum take-off weight.

MZFW Maximum zero fuel weight.

NASC National Aviation Security Committee.

NATS National Air Traffic Services.

nm Nautical miles.

NOC British Airways' Network Operations Centre.

NPR Noise preferential route.

OFZ Obstruction-free zone.

OMC Operational Monitoring Centre.

PAPI Precision approach path indicators.

PCN Pavement classification number.

PRT Personal rapid transport system.

PSZ Public Safety Zone.

Push-back The process of pushing an aircraft backwards away from an airport gate by means of a low-profile tractor or tug.

RDT Radio data terminal.

RET Rapid exit taxiway.

RIMCAS Runway incursion monitoring and collision avoidance system.

Runway designator A painted letter at the start of each runway to help identify it.

SALS Supplementary approach lighting system.

SAMOS Semi-Automated Met Observation System.

SAU Stand Allocation Unit.

SEG Stand entry guidance.

Shoulders Concrete areas to the sides of the contact lines on a runway to prevent erosion by overhanging engines.

SID Standard instrument departure.

SIS Staff Information System.

SMR Surface movement radar.

SNIB Stand number indicator board.

SOC Senior Operations Controller.

STA Scheduled time of arrival.

Stack or **hold** Area of airspace in which aircraft approaching an airport can be held in an orbital pattern pending a landing slot becoming available.

Stand Aircraft parking area.

STAR Standard Terminal Arrival Routes.

STD Scheduled time of departure.

STL Service Team Leader.

Stopway A rectangular area at the end of the available take-off run.

SWDS Surface wind display systems.

Taxiway Route used by aircraft moving to and from the runway to their allocated parking stand.

Taxiway strip and graded area A designated area clear of potential obstructions on either side of a taxiway.

TC Terminal Control.

TCC Terminal Control Centre.

TDM Terminal Duty Manager.

TDZ Touchdown zone.

TEAM Tactically enhanced arrivals measures.

Threshold The commencement of the runway area available and suitable for the landing of aircraft. The markings that indicate its start are sometimes referred to as 'piano keys'.

TMA Terminal Manoeuvring Area.

TOCS Take-off climb surface.

TODA Take-off distance available.

TORA Take-off run available.

Touchdown zone markers Parallel stripes that indicate the area on which pilots should put the wheels of their aircraft when landing.

Track miles The distance an aircraft must travel on a standard routing to approach the runway rather than the aircraft's actual physical distance from the airport.

UCAS Uncontrolled Airspace.

UKOP UK Oil Pipeline system.

ULD Unit load device.

VCR Visual Control Room.

VOR VHF omnidirectional radio range.

Wheel track The width of an aircraft's main undercarriage.

WHIRS Wider Heathrow Integrated Rail Strategy.

WIB Western Interface Building.

The Flight Path to Growth

A key driver of economic growth and a catalyst to international trade and tourism, the aviation industry generates $2.2 trillion worth of business and supports over 57 million jobs in today's global economy. These jobs provide challenging career openings in an industry that delivers benefits that extend beyond borders.

Our industry was built with the skills of talented people who were passionate about connecting the world. To continue to grow and thrive, airline management and individual airline professionals need to continuously make the right investment decisions with regards to expanding and reinforcing relevant skills and knowledge.

Developing Human Capital for Tomorrow's Air Transport Industry

IATA's Training and Development Institute (ITDI) is focused on meeting the air transport industry's need for flexible, high-quality training to develop human capital for tomorrow's air transport industry. As the leading provider of global aviation training solutions and professional development programs, we have an unparalleled track record in providing expert training for all critical areas of the aviation business. Moreover, we have partnered with some of the world's leading educational establishments to augment our general management offerings and ensure that courses are delivered with the latest learning techniques. Comprising the widest possible range of course material with a strong geographical and cultural reach, ITDI training provides rewarding opportunities for all aviation professionals.

Offering Internationally Recognized Programs

Our internationally recognized programs reflect the changing needs of our industry and cover all areas of aviation including: airlines, airports, ground services, civil aviation authorities, and air navigation services. At ITDI, courses are developed and taught by industry experts who engage participants through case studies and simulation exercises. Participants study from IATA training materials that are produced in-house, by the very organization that creates the industry guides and manuals you rely so heavily on every day.

Dedicated to Helping You Achieve Your Goals

ITDI offers dynamic and innovative training solutions with the global reach to positively impact every level of the aviation supply chain. We harness the power of technology to bring you a wide variety of training options that are innovative, flexible and cost-effective. Simply put, it is our mission to educate the industry while helping you to achieve your personal and professional goals. To that end, we continue to build partnerships with esteemed industry and academic institutions to offer a host of MBAs, diplomas and certificates so that you can earn industry-wide recognition for your academic achievement and reach your true potential.

www.iata.org/training

Index